Worm Farm Management

Worm Farm Management

Practices • Principles • Procedures

ERIC WILSON

Kangaroo Press

WORM FARM MANAGEMENT

First published in Australia in 1999 by Kangaroo Press
an imprint of Simon & Schuster (Australia) Pty Limited
20 Barcoo Street, East Roseville NSW 2069

A Viacom Company
Sydney New York London Toronto Tokyo Singapore

National Library of Australia
Cataloguing-in-Publication data

Wilson, Eric. 1951–
Worm farm management: practices, principles, procedures
Includes index.

ISBN : 0 86417 986 3

1. Earthworm culture. 2. Earthworms - Marketing. I. Title

639.75

Set in Palatino 10/13
Printed by Kyodo Printing Co. Pte Ltd, Singapore

10 9 8 7 6 5 4 3 2

Dedication

I dedicate this book to my family without whose existence there would be no reason for being.

To my mother and father, Betty and Eric, for their exceptional patience and understanding during the five years I grew worms commercially on their property while I struggled to make ends meet and learned the tricks of the trade. My appreciation of your combined strength in the face of adversity and the profound effect it has had on my life can never be measured. I love you.

To the memory of my sister Lorraine, whose life was terminated by an act of violence in the prime of her youth. You are constantly on my mind. I draw upon a whole host of emotions in your memory to get through life. I love you.

To my children Samantha, Talbot and Kirsty, whom I adore and admire, and my grandson Blake, for the joy of being part of the family unit and for the encouragement, support and love that has enabled me to finally write this book. I love you.

Acknowledgments

My thanks go to Robert and Sally Howlett for supplying photographs of their worm farm, and to Diane Tatum, John Vickery, Adam Loel, Roy Ruckman, Pam McKinnon and Workin' Worms for allowing me to use photographs of their farms. All photographs not acknowledged are by the author.

Contents

Introduction

This book is about the compost worm, a term which includes those species generally referred to as the red, the tiger and the Aussie blue compost worm. These species (as a combination or as one species only) can successfully be commercially bred and grown in your backyard. Big is not necessarily better, particularly with an intensive livestock industry. Anyone can be a successful worm grower—all it takes is a little common sense, an understanding of the worm's natural behaviour and a rotational system of management.

You will notice an underlying theme with constant repetition throughout the book of such terminology as 'stocking density', 'surface area', 'temperature', 'working layer', 'castings', 'husbandry management', 'management systems'. This book is about the practices, principles and procedures of worm husbandry management, from a non-technical point of view, and as such requires the constant interaction of the variables that constitute these terminologies.

All information is based on the experiences and observations resulting from my eight years of worm growing. The explanations I give are based not on scientific evidence or trials but rather are my conclusions based on those eight years' observation and experience. No part of this book is taken from other people's research.

I believe I empathise with and know worm behaviour—I have tailored my management systems to accord with my belief. I hope this book sheds some light for you on the behaviour of worms, and on what is required from a husbandry and management point of view.

My systems, my observations, my explanations are exactly that—*my* experiences only. This does not necessarily mean that you will agree with me or that my system will work for you. It is up to you, as your own manager of your own business, to discover what works best for you.

I write this book in my own style, with whatever Australian colloquialisms are part of my personality.

I am not a biologist, thus this book is not about the biology of the worm.

I am not a market analyst nor a psychic, thus this book is not about where and how to find your markets, how long the industry will last or what direction it will take.

This book is about worm husbandry management, its practices, principles and procedures.

I wish your venture success.

Eric Wilson
Director, The Australian Worm Growers School Pty Ltd
Galston 1999

CHAPTER I

Setting Up

Overview

The worm industry is an intensive industry which does not require broad acreage to operate on. Worm farming can be a successful home industry; depending on your scale of economy it requires only minimal amounts of space, time and capital outlay.

What are the advantages of growing worms?

- Worms consume your organic garbage, thus saving on waste disposal costs.
- Worms convert your organic garbage into valuable worm manure (castings), nature's perfect blend of plant food, thus you save on buying fertiliser.
- Worms and their by-products can be sold to make additional income.

What is needed to start a worm farm?

It is prudent to establish one nucleus bed to begin with so you can observe the nature of the beast before establishing an organised farm and management system.

To get started you need:

- An above-ground bed
- A quantity of worms to suit your bed size
- Manure supply
- Tools of the trade.

What size bed and how many worms?

The dimensions of the bed should suit the number of worms (stocking density) that you intend to begin with (your nucleus), e.g. a 2.1 m x 1.2 m bed = a surface area of 2.5 m^2. Allowing for a maximum of 2 kg breeding worms per m^2, you would require a maximum of 5 kg worms for this space.

How much manure is needed?

Manure can be acquired from any animal source but should be freely available and accessible. As a rule of thumb, quantity should initially be based on worms consuming their own body weight per day, e.g. 5 kg of worms will require 5 kg of food per day.

What tools are required?

You need only the most basic garden tools (shovel, fork, wheelbarrow and water hose).

What production can be achieved?

- Documentation shows that under ideal conditions, as a rule of thumb, one worm can lay an egg every seven days, i.e. 52 eggs per year.
- The average hatching from one egg is four worms.
- Under ideal conditions, one worm can produce over 200 offspring per year. This would constitute a 200 times return on your original stock numbers, i.e. a ratio of 1:200.
- If your stock only produced one egg per year per worm, this would still constitute a four times return on your original stock numbers, i.e. a ratio of 1:4.

Equating these statistics to a $ value

- One worm lays 1 egg per year = 4 offspring = ratio 1:4, or 1 kg worms produces 4 kg worms per year.
- One worm lays 5 eggs per year = 20 offspring = ratio 1:20, or 1 kg worms produces 20 kg worms per year.
- One worm lays 10 eggs per year = 40 offspring = ratio 1:40, or 1 kg worms produces 40 kg worms per year.
- Using these figures and the going rate for a kilogram of worms, you can calculate the value of your production.
- Worms are livestock. To achieve effective breeding rates they require good husbandry management techniques.

The question is: *How good a manager can you be?*

Production costs

How much does it cost to produce one kilogram of worms?

You, as a business manager, have to know what your cost of production is and where your break-even point is.

Cost of production includes all expenses in the operation of your business, including set-up or establishment cost (a fair and equitable time period must be allowed to calculate establishment costs, back to a weekly or monthly basis).

Your labour is an important ingredient and must also be calculated into the equation.

Production is about producing on a regular consistent basis.

If your market is the live worm market and your worms are not breeding then you are not producing, but it is still costing you for feed, water, electricity, labour, advertising, and general running around, to look after or maintain your enterprise. This is why it is so important to have a business plan which incorporates a sound husbandry and operational management plan that will allow you to see at a glance where a problem is located.

For a minority of inefficient producers, the bottom line at the end of the day may be to 'close the doors and go fishing'.

Worm husbandry entails a combination of management skills based on normal agricultural animal husbandry management, practices, principles and procedures.

It is important to know the nature of the beast if you intend to control and profit from it.

General guidelines for bed set-up and design

The breeding/incubation/growing bed is the foundation of your enterprise. Finding the balance between size and design which is best for your stock and most convenient for you, from the physical point of view, is important.

The human requirement is that all beds be easy and convenient to manage and maintain. The stock requirement is the positioning that best suits the requirements of breeding and growing. The successful set-up is designed to capitalise on the best usage of space so that your venture is not only accessible, practical and rotational, but also neat and tidy.

Particular attention should be paid to your estimated market; e.g. if your market frequently requires stock quantities of approximately 25 kg, it may be prudent to have beds with a surface area of 2.5 m^2 which have a holding capacity of 25 kg of worms at marketable size. A bed measuring 2.1 m x 1.2 m would provide the required area.

Shape and size

There is no hard and fast rule concerning the physical shape of your beds, providing that the total surface area obtained matches the desired stocking density.

Height

On-ground beds (positioned on the soil surface) On-ground beds should be constructed to a height of approximately 450–500 mm (18–20 in). This is more beneficial for your back, as the top of the bed is about knee height and acts as a pivot point when working around your beds, e.g. tossing. This height also allows you to kneel on the ground and lean over comfortably to work. A height greater than this is not only hard on the thigh muscles but makes it very awkward to reach and work the bottom. A height less than this is damaging to the shinbone and gives no pivot point for balance.

Off-ground beds (elevated sufficiently to be off the soil surface) A drainage allowance has to be made for the height of the bearers and joists necessary to carry the weight of bedding. Low-level off-ground beds work at the expense of user comfort. See Plates 2 and 3.

Raised beds (elevated sufficiently for user convenience) The working height should suit the user, so that minimal back bending is required. A common height is approximately 900 mm (35 in). See Plate 4.

Width

It is important that you be able to comfortably reach the middle from both sides, which dictates a maximum width of approximately 1.6 m (5 ft 3 in). If the bed's access and location restrict the operator to working from only one side, the effective convenient working width is reduced to 800–900 mm (30–35 in).

Length

Length is the variable factor which will determine surface area. Smaller beds are more convenient and easier to handle than larger ones.

Materials

Whether your beds are constructed of timber, brick or corrugated iron, or any other material, is of little consequence as long as the material you choose has the physical characteristics to retain your stock and bedding in the one location. The availability of suitable building products and their cost will probably be the determining factor.

A plastic-lined timber bed that can be easily moved by two men is ideal. The construction is simple, materials are readily available and everyone has access to a hammer and saw.

Construction

A basic four-sided structure will retain the bedding for the worms, but it is also necessary to build the bed so that your stock cannot escape—you need to know that your stock are where you left them.

Compost worms by their nature will not burrow further than a few millimetres into the soil to provide covering for survival. In a growing/breeding bed they will restrict themselves to their rich organic food and their own castings.

The worm's basic instinct is to be a surface feeder—they will always come to the top and once on the surface (if conditions are right) they will wander.

An on-ground bed should be constructed so that it is dug into the ground for about 50 mm (2 in). If you plastic-line the bed it is an easy process to sit the timber frame on the ground and dig a trench to bury the plastic. This prevents the worms exiting under the timber when bedding volume is low.

Plastic up the sides of the beds not only protects the timber from moisture and rotting but also ensures that the worms cannot get out through the spaces between the slats and have only one escape route, i.e. up the side and over the top.

The addition of an escape-proof lip, the use of covers or low voltage lighting, or a combination of these, with the necessary husbandry techniques, will prevent your worms from exiting their beds.

Internal drainage A very simple and effective method of drainage within a bed. This bed is totally plastic-enclosed in 200-micron builder's plastic, with the only avenue of escape (up the sides and over the top) prevented by the escape-proof lip. The base plastic in this case was added as an afterthought to combat long-reaching tree roots, and was taped to the existing side plastic with duct tape. Furrows were gouged into the soil with the plastic contoured into them to take the agricultural drainage pipes. The earth under the plastic was also contoured to assist the flow of water towards the pipes. The ag pipe is fitted with a filter sock which is essential for its effective operation (without the sock, fine sediments would block the pipe very quickly). The end of the pipe is simply pushed through a hole in the plastic about half the actual diameter of the pipe. The flexible nature of the builder's plastic allows it to stretch without tearing, then shrink back to form a tight seal around the ag pipe.

The escape-proof lip

Worms can and will migrate in mass numbers when the conditions are favourable. Without fences to contain sheep and cattle, you would not be able to find your stock to be able to manage them—worms are no different. Without lighting to keep them suppressed, it is essential to design your beds with an escape-proof lip.

Worms do not like a dry surface, particularly one that is rough and non-adhesive, like timber. This essential feature begins with a piece of timber (approximately 150 mm x 25 mm or 6 x 1 in) laid horizontally on top of the edge of the bed all the way around the perimeter, like a seat all the way around. It is important that the major part of the timber lies over the bed itself (the wider this part the better, 100–150 mm or 4–6 in).

This 'seat' must be cushioned with Silastic or a similar silicon compound and coach-screwed or nailed tight to the top vertical plank. At the points where the various planes meet, it is also important to apply a silicon compound that expands and shrinks in order to fill the gaps and prevent water leakage.

You now add a return (i.e. the lip) to stop water dribbling or being blown back under the 'seat'. This lip is a piece of timber (approximately 50 mm x 25 mm or 2 in x 1 in) placed vertically on the inside edge of the seat. The top of the lip is flush with the upper surface of the seat. This piece of timber must also be cushioned with a silicon compound, firmly nailed and go right around the internal perimeter of the seat.

The timber, being a porous substance with knots, cracks and fissures, must now be treated with a sealant or water-repellent product, e.g. lacquer, to prevent any seepage or leakage that would allow the undersurface to become wet.

To ensure the continual effectiveness of an escape-proof dry undersurface, this lacquered layer and siliconed joints must be maintained. See Plate 5 for a beautifully made escape-proof lip.

Location and drainage

Beds can be situated in full sun, in filtered light, in sheltered areas or in sheds but particular attention should be paid to your chosen location to facilitate drainage.

Under a rotational system of management, the ideal would be to have a combination of locations and conditions to suit the seasons—full sunlight for winter breeding (beds in the open), filtered sunlight for spring and autumn breeding (beds under the shade of trees—see Plate 1) and sheltered shaded conditions for summer breeding (beds under shade cover or in a shed). Not everyone can manage this, of course.

To capitalise on the maximum amount of sun, the beds should run north/south—in this way, in winter, for example, only a small section is constantly shaded. Keep in mind also that positioning and direction are relevant to drainage and take account of the lay of the land.

Heavy clay areas with poor absorption may need additional drainage or may require the construction of off-ground beds. See Plate 2.

Setting up beds under trees with a voracious rooting system, e.g. pines and gum trees, will require plastic lined on-ground beds with internal drainage, or the construction of above-ground beds.

There is no such thing as a bad location if you go about things in a well-thought-out manner, as your entire set-up should be designed around manageability and practicality.

Coverings

Internal covers

Not every grower chooses to cover the surface of the beds, but the use of carpet, hessian or other coverings can be a sensible option to reduce evaporation, eliminate possible bird problems, protect exposed eggs from the elements and seal the surface to provide a comfort layer for your stock.

External covers

Any bed can flood very easily during heavy downpours of rain, even with good drainage, therefore take preventative action by having adequate covers on hand.

An A-frame structure over the bed with fold-back or roll-up facilities for your covering is prudent and practical. The height to the apex of the A-frame should be enough to facilitate run-off and adequate air ventilation to the surface of the bed.

The A-frame can have the dual role of providing shade with the use of shadecloth, which allows water and air to penetrate, or be primarily a sun, wind and rain block if covered with tarpaulin. A system that incorporated both shadecloth (on the inside) and tarpaulin (on the outside), although expensive and fiddly, would be ideal.

The A-frame also gives a physical structure allowing for the possible attachment of poly-irrigation hoses for a sprinkler/mister system.

Regardless of possible heavy downpours of rain, the provision of external covers for exposed beds can be most beneficial for providing shade against intense summer sun and shelter against winter winds (stops the wind-chill effect).

If there is no provision for a permanent external cover, temporary protection against anticipated heavy downpours may be provided by a sheet of corrugated iron or fibreglass or similar laid over the bed and held down with counterweights, such as bricks, stones or timber.

Power supply

Worms migrate when nature's conditions are favourable (dark and wet). It is advisable to position your set-up within reach of electricity to supply lighting. Lighting requirements to prevent walkouts are minimal—a 12-volt garden lighting system positioned within the bed will suffice. The lighting must be placed to eliminate shadows if it is to be effective.

Given the dangers inherent in mixing water and electricity it is probably more prudent to construct outside beds with an escape-proof lip.

Basic tools of the trade

No matter what your scale of economy, the basic tools are still required.

Shovel A short handle with a medium dished face is ideal. Used for moving and mixing your primary food source and to feed your stock. Do not use to aerate the beds.

Pitchfork A light 3-fingered spring steel pitchfork is ideal. Used for aerating and tossing your beds.

There will always be damage done to the worms (called 'acceptable losses'), either through spearing them directly with the pitchfork or by injuring them during the lifting and tossing of the bed materials. This is why a pitchfork with fine points and thin smooth fingers is important to minimise damage and why it is important to *maintain* a loose, friable working layer.

Garden fork A normal 3-pronged garden fork with a short handle is most useful for removing the working layer or baited sections and to aerate and turn manures.

The fork can be used for aerating your beds if necessary but becomes heavy to use and damages a lot of worms.

Rake A multi-function tool with a variety of uses, including breaking up castings to a fine tilth, bringing all the material to one point for extraction in a bed after all the worms have been harvested, and levelling or evening out manures once the beds have been blanket fed.

Wheelbarrow Choose a large builder's barrow with good balance, good grip and a pneumatic tyre. Use it to transport manures and castings, to clean out the beds, to mix supplementary foods, to blend manures, to soak products, etc. The wheelbarrow is a versatile labour-saving implement that you cannot do without and that you will use every day.

Hose and fittings The normal garden hose that suits your water pressure and the appropriate fittings for spray adjustment are all that is required. The hose should be long enough to reach all points of watering without strain.

Sorting table The minimum size advisable for a sorting table is 2.4 m x 1.2 m (8 ft x 3.5 ft), with a height of 900 mm (35 in); a smooth impervious top is necessary (plastic overlay would suffice). The table is used for light-extraction; this minimum width allows you to work from both sides.

Although a table is the most convenient, the back of a tabletop truck or a sheet of plastic on a flat surface will also do the job.

Scales The weighing capacity of your scales should match a manageable weight of worms. Worms in 1 kg (2.2 lb) lots are a common manageable size so buy scales which measure to no more than, say, 1.2 kg (2.6 lb). This allows the full working use of the spring mechanisms and gives an accurate reading.

Buckets and boxes It is always handy to have a selection of plastic buckets with handles and lids, from 10–20 litres in size, for convenient measuring, storing and carrying.

Polystyrene boxes are ideal for storing moist products like cocopeat, packaging material and castings—they insulate the contents and prevent them from drying out, and because of their design they stack flat and square—most convenient.

Assorted extras These can include watering cans, pH testers, gumboots, etc.

Supplementary machinery As the scale of economy increases, your requirements for the machinery necessary to operate efficiently also increase.

- Mulcher/muncher—for processing putrescible wastes, etc
- Hammer mill—for processing grains, straw, paper, etc
- Cement mixer—for mixing supplementary foods, packaging materials, blending, etc
- Rotary hoe—for the preparation of loose friable manures and as an aid in the composting process
- Front-end loader or bobcat—for turning composting manures, moving large amounts of material easily, feeding the beds, emptying the beds, etc
- Pumps and irrigation equipment—for convenience and ease of watering
- Harvesters—for a mechanical method of worm extraction: sieving, screening and the removal of eggs and adolescents
- Sealer—for sealing plastic bags of castings ready for the market.

Aerobic digester

An aerobic digester is a beautiful multi-functional addition to a 'closed system' (no leaking of contents into the substrate) to complete the loop. It is simply a tank in the ground where all the run-off (or leachate), the result of watering, is collected. The tank or receptacle must be large enough to cope with the anticipated run-off created during the maximum watering periods of summer. See photo below.

Logically, your worm farm's set-up will suit the lay of your land, thus the tank will be at the lowest point to facilitate drainage through the beds and the downhill flow of waste water to the tank.

A filter screen is advisable to prevent rubbish and pollutants from outside sources entering the tank (leaves, frogs, grasshoppers and so on).

Inside the tank is a submersible pump with a return so that the water is constantly agitated and moving and absorbing oxygen, keeping conditions within the tank always aerobic.

The aerobic digester is your breeding system for microbes. The fluid draining from the worm beds contains millions of microbes. All you have to do is feed them, preferably with the same manure source and micro-

Aerobic digestor This 1200-litre plastic tank has been dug into the ground and receives drainage from five separate beds. It contains an electric centrifugal submersible pump (suspended by a synthetic rope) and the provision for fresh water (poly pipe). The light discolouration on the black tank is calcium deposits from bore water. The tank is counter-weighted with blue metal and the pump positioned at a height to prevent the tank from popping out of the ground as the water table rises after rain.

organism diversity which you feed the worms. This will ensure that the water from the digester, when watered back across the beds, assists in the breakdown of primary food sources. Microbes provide exceptionally good nutrition for the worms—the more you have the better the food source and the better the environment of the worm beds.

The size of the tank will determine how much food you need to put in it, but certainly no more than half a shovel of manure or, say, 5 litres of pollard or 2 litres of molasses to a 1000-litre tank. Let the smell and froth that sits on top of the water the next morning, i.e. 12 hours later, be your guide to the level of microbial action and frequency of feeding the system (not the worms). Molasses is very easy to use and the microbes respond very quickly. The next morning, you have a sweet sickly frothy broth that is literally alive with micro-organisms.

Micro-organisms 'peak' very quickly and when their food supply disappears they die very quickly. If the aerobic pump is turned off the system very quickly consumes whatever oxygen is in the water and then begins to work in reverse, i.e. anaerobically, and within two days you will have a disgusting smelly septic system. If this happens, simply turn the pump on

Mechanical help This handy 20 hp Kubota tractor (with 4-wheel drive and power steering) is fitted with a front-end hydraulic bucket and 3-point linkage PTO-driven rotary hoe. A versatile piece of machinery such as this, which allows the attachment of a range of implements, is of paramount importance once you reach a scale of operation sufficient to justify mechanisation. This picture shows a windrow of composting manures that has been rotary-hoed, tossed and oxygenated—the final preparation prior to feeding.

and either empty the contents out and start again with fresh water, or activate and rejuvenate the system by adding oxygen, a little food and maybe some pelletised freeze-dried micro-organisms for faster action. Within 24 hours the system will be aerobic once again.

pH levels and watering

Having the correct pH level in the worm beds is most important for breeding. By adjusting pH back to neutral, you also free up nutrients that were previously unavailable and in the process transform the bedding structure to an easy workable friable state. An easy way of doing this is with the use of hydrochloric acid.

Simply add hydrochloric acid to your watering tank or aerobic digestor until it reaches a pH of 5.5–6.0 (it is essential that you use a water pH tester, not a soil pH tester). This method is very economical, instantaneous and simple (avoid breathing the vapours and protect against spillage on your skin) and in the process of watering you are constantly changing the pH.

With a closed system it is easy to know the overall pH of the beds simply by doing a leachate pH water test at the point where the leachate exits the bed. Prior to the leachate entering the tank, take a sample and test it—this, of course, will give you a different reading to what you will get from a sample of bedding from the working layer (because of the food characteristics, not the casting characteristics).

The sensible management of molasses (which is acidic) can eliminate the need for using hydrochloric acid; however, for a quick adjustment, hydrochloric acid should be considered.

Using an aerobic digester will give you an environmentally safe closed system (with no leaching) that collects the run-off and provides a treated product that can be re-watered over and over again to keep the beds moist, at the same time liquid feeding the worms and their environment with microbes and constantly adjusting the pH.

Introducing stock

In the setting up of a new breeder bed, it is important that your stock acclimatises and accepts the new environment as quickly as possible. Once the worms are safe, secure and settled you can begin normal husbandry practices.

Signs of distress include balling up, dispersal and disorientation, or simply showing signs of not wanting to move out of the packaging material.

The breeding of worms is largely about surface area and stocking density (i.e. bed size should equate to the amount of worms or vice versa) and with this in mind you now have two considerations:

Access to casting or existing bedding

Casting or existing bedding contains the environmental properties with which the worms are familiar so there should not be a problem with the introduction of your stock.

Before introducing stock, prepare the bed so that its structure is granulated for ease of penetration and its shape is duned, with approximately 150–225 mm (6–9 in) depth in the centre tapering down to approximately 50 mm (2 in) depth on the periphery—this allows for variation in temperature, drainage and moisture.

Spread your worms over the entire surface area and lightly 'massage' or 'tickle the surface' for even distribution.

Worms should be introduced to a new bed only when the bed is cool, preferably early in the morning or in the late evening, with sufficient light to force them off the surface and into their new home. If the bed is hot they will become confused and probably will not enter.

Once they have disappeared, give the bed a good watering (without destroying the structure) and cover if required.

Access only to primary food source

Release the worms (in their transporting/packaging material) in a row running from end to end in the middle of the bed.

Place your chosen manure on both sides of this row so that it is in contact with this row of worms. The pile of manure should be about the same size as the worm/bedding row. Give the rows of worms and manure a good watering and cover with hessian or carpet. The purpose of this exercise is to allow, not force, the worms to move sideways into their new manure source at their own rate.

The next day, check to see whether the worms have entered the manure; if they have not, allow one more day, then consider the removal of this feed as there may be a problem with it and try another batch.

Once the worms have entered their new source, mix the three rows into one. This then constitutes the basis of the bedding.

It is important to build up this row to 150–225 mm (6–9 in) height in the middle as quickly as possible, tapering the edges down to touch the sides of the bed—this 'dune' effect should be maintained as the bed builds in height as it allows for variations in temperature, drainage and moisture.

Horse manure is very good in this initial set-up because it is loose and friable and the worms very quickly enter and spread over the entire surface area.

CHAPTER 2

Worm Biology

Physiology

The most basic description of a worm is a sliver of protein that eats voraciously (up to its own body weight per day), excretes copiously and reproduces at an amazing rate.

More precisely (but not in detail) the smooth body of the worm is made up of rings called annuli. Its basic structure is quite simply described as two tubes, one inside the other—the outer one being the muscular body wall, the inner one a digestive tube. There are no eyes or ears, but they are sensitive to heat, light and touch. The worm breathes through its skin, taking in air from between particles of soil. It feeds on dead organic matter in or on the soil.

The worm moves by lengthening the front part of the body and pushing through its surroundings, then drawing up the hind part. Minute bristles (setae) prevent the worm slipping.

All worms are hermaphroditic, having both male and female sexual organs, but must mate with another worm, each providing the other with sperm to fertilise its eggs. The fertilised eggs are laid in an egg capsule secreted by the clitellum ('saddle'), a group of distinctive body rings some distance behind the head. The capsule is wriggled along the body and over the worm's head, closing completely to protect the contents. The young emerge from the capsule after several weeks.

Clitellum The presence of the clitellum, essential for the mating act, is an indication that the worm is sexually mature. On the red or tiger worms, the band can be quite prominent, and the worms actually slide under the secretions of each other's clitellum to 'knot'—see Plate 7. On the Aussie blue, the clitellum is not prominent but is distinctive in colour. Aussie blues do not knot but sidle alongside each other to exchange ova and spermatozoa.

Once the worm is sexually mature, the clitellum is always present but is not necessarily pronounced, and comes and goes with the worm's condition. A large clitellum is a sign of a worm's 'coming into heat', i.e. primed hormonally for the act of reproduction.

Breeding cycle

The breeding cycle begins on Day 1, when the first worms are turned into the breeding bed. Observation is the key. Breeding will take place according to Nature's Equation, my term for the combination of balances between soil and atmospheric temperature, pH, moisture, food supply, oxygenation, stocking density, surface area, bed volume and so on.

You must continually observe the conditions of your beds and vary those conditions until you are content with the level of breeding you wish to maintain. Optimal breeding begins, in the first instance, with the right stocking density, i.e. 1–2 kg per m^2 (3–6 oz per ft^2). This stocking density should be maintained.

Breeding management will vary depending on whether you choose to use traditional above-ground beds, windrows or the 21-day cycle.

Temperature and the seasons—the breeding cycle nature's way

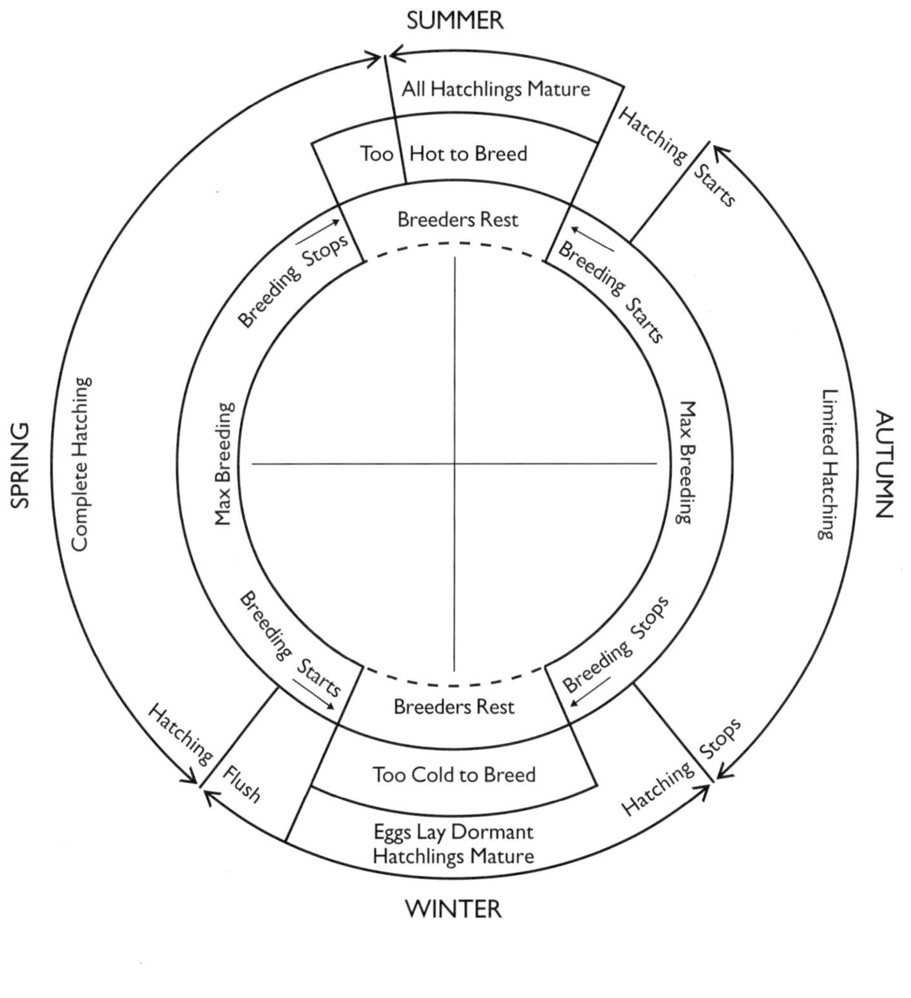

Eggs

Freshly laid eggs are a translucent white or frosty colour that quickly changes to a yellow/green hue. As the days pass, they turn light brown and then to a darker brown-blackish colour prior to hatching. See Plate 8.

The egg has a semi-permeable membrane that can quickly dry out so it is important, after tossing beds, to make sure that the surface is covered with a moist layer of bags, carpet or similar to ensure protection against wind and sun. The semi-permeable membrane regulates the transfusion of waste gases and maintains the moisture content under good conditions. It is not unreasonable to assume that a certain amount of absorption occurs via osmosis through this membrane to trigger a chemical signal indicating an environment with the favourable conditions of food and moisture supply to hasten the hatching process.

The hatching process is, however, primarily controlled by a constant temperature, significantly higher than that required for breeding. The beds should literally be lukewarm to touch.

The size of the egg capsule is in direct proportion to the diameter of the worm—a small worm lays a small egg, a large worm lays a large egg. The eggs are literally rolled off the head by a muscular process similar to peristalsis and are laid where conditions at that particular time best suit their survival.

The number of hatchlings per capsule appears to depend on the period of time from lay to hatching. Under ideal conditions, eggs will hatch after 21 days, when it is not uncommon to have 6–8 worms emerge from each capsule. Under normal conditions in a normal bed with varying temperatures the average is 3–4 hatchlings. After a prolonged period of time, e.g. lying dormant through winter, or under variable climatic conditions, it is usually a case of one large worm emerging from each egg. This relates to the finite food supply within the egg capsule and the survival of the fittest.

A range of colours in the eggs in the breeding beds at any one time indicates continual breeding. The number of new-laid eggs should be in proportion to the number of worms in that bed. Under ideal conditions each worm can lay an egg every seven days, subject to Nature's Equation, thus if you have 200 worms in a breeding tray you should theoretically get 200 eggs per week—if you only have 10 or 20 then something is wrong.

For ease of management, breeding beds should be set up with minimum depth (subject to time of year) and breeders should be evenly distributed over the entire surface area. In this case it is easy to do an egg estimate simply by extracting an entire section, say one square metre. Estimate the weight of the sample, then mix it thoroughly to evenly disperse the eggs. Take a small sample, weigh it, wash it through a sieve and count the number of eggs in the sieve. From this count, work backwards by multiplying to get a representative egg count for the bed (you should be within 10 per cent of the actual).

Newly hatched worms are generally pinkish in colour. If they are not (which happens with large numbers emerging from one egg or hatching from a small egg—where they will be white) they change colour very quickly. Light coloured hatchlings should not be confused with the yeast worm which is white or frosty in colour and grows to about 10 mm (0.4 in).

Breeding flush

A breeding flush can occur at any time of the year with the duplication of spring conditions. It may only last for one or two weeks; then as quickly as the flush has started, it is finished. This situation is totally different from the intermittent breeding that goes on all the time. It is truly magic and can clearly be seen in beds exposed to direct sunlight in winter with the coincidence of a warm sunny week.

Hatching flush

A hatching flush is directly related to a breeding flush, or where you have an accumulation of eggs lying dormant. When the conditions are right, you will have an incredible number of hatchlings appearing in a very short time, as opposed to intermittent hatchlings resulting from intermittent breeding.

CHAPTER 3

Bedding Systems

Every business has a management or operational plan. Worm growing is no different from any other livestock industry, where the accepted practice is to isolate breeding stock from growing stock. To do this, you need to operate two distinctly different bedding systems: breeding beds and growing beds, within either a traditional above-ground system, a broad-acre windrow system.

Traditional above-ground beds (on-ground, off-ground and raised beds) These are manageable, containable, neat and tidy.

Windrow Broad-acre usage—cheap, less easily managed, machinery and earthworks required.

21-day cycle Maximum utilisation of space, requiring work discipline and time management, and consistent stock control.

The majority of worm enterprises operate with the traditional on-ground beds. However, your particular scale of economy, economics and available space may influence you to consider windrows, off-ground beds or the 21-day cycle.

Breeding beds

Breeding beds are set up with the conditions that most favour egg production and hatching. See Plate 16. Don't sell out of your breeding beds—you need these producers—and ensure you extract adolescent worms on a regular basis.

As it is impossible to remove all adolescents, periodically test sample breeding beds to maintain stocking density.

To obtain the best results from breeding stock, maintain the bedding following this checklist:

- The optimum stocking density is 1–2 kg per m^2 (3–6 oz per ft^2)
- Stock are in store condition
- Stock are on a rising plane of nutrition
- pH is neutral to slightly alkaline
- Bedding is aerated in the top 100–150 mm (4–6 in)

- Bedding is loose and friable
- Bedding has a diversity of insect life
- Bedding has sufficient volume for temperature and moisture variation
- Feed and aerate to suit breeding flushes and climatic conditions (cold nights and warm days).

Growing beds

Growing beds are set up to grow adolescents to a marketable size—not to breed more worms. The ultimate aim is to sell the minimum number of worms possible per kilogram while maintaining a high stocking density (10 kg per m^2 or 32 oz per ft^2) to allow easy extraction for harvesting. Keep the bedding very moist—but not sloppy.

Feed plenty of good quality food as these youngsters have a voracious appetite. Manures from ruminants whose digestion is more complete, e.g. cattle and sheep, are a safe choice for a primary food source, and combined with high protein supplementary foods such as pollard or chook pellets, give good results. There is no need to adjust pH in the growing beds.

Keep the surface bubbling and alive with worms which are being pushed to their genetic potential.

Harvest the entire bed when ready for marketing.

Rob's grower beds These on-ground grower beds are approximately 4 m long x 1 m wide (13 ft x 3.3 ft). They are constructed from a roofing/wall cladding colour-bonded sheet metal product. The deep-rib profile of this cladding gives exceptional strength. The sides are fixed to corner posts by wood screws. The roof frame, a simple post and rail affair, provides support for a throw-over tarpaulin which is fixed in the middle and tensioned with elastic ropes. (Courtesy Robert and Sally Howlett)

High stocking density/low breeding rate

As breeding numbers increase so too does stocking density, and consequently the breeding rate will slow down. What options are available?

Splitting the beds

This, of course, is the easy answer—empty out half the contents (including castings and worms) from one bed and put it into another bed.

Thinning

Thinning involves the removal of worms only (via light-extraction or mechanical harvesting) to reduce the stocking density and either placing them in a new bed or packaging them for sale.

Weaning

Under this procedure, you extract the adolescents and empty them into a separate growing bed. The management of this bed is, of course, different from that of a breeding bed. In this situation you have masses of adolescents growing very quickly, with a voracious appetite for good quality food. Weaning can be accomplished using one or more of the following techniques:

Internal weaning tray Leave the tray, which has been baited, in the bed, thus luring the adolescents into the tray.

External weaning tray Place raw worms (naked worms, in mass, out of their environment and without bedding) in the tray and allow the adolescents to move out into the bedding.

Weaning fabric 90 per cent shadecloth works well. Place fabric on the surface of the bed with the bait on top, thus encouraging the adolescents to move through the fabric into the bait.

Mechanical harvesting When the bedding is at the right consistency, the majority of the adolescents will fall through the mesh. They can be collected and placed into a growing bed.

Windrows

This term is derived from the agricultural practice of turning mowed hay into a row so it will dry with the assistance of the wind and sun. This is the least easily managed of all the systems but it is very economical to set up.

The windrow simply entails a row of bedding or manure placed either directly on the ground or on a sheet of builder's plastic, weed mat or similar product. To this you then add your worms and then apply the normal worm husbandry practices of watering and feeding. Windrows are a different kettle of fish to the traditional above-ground breeding beds and should be set up with two important factors in mind: aeration and watering/covering. See Plate 9.

Ideally, the windrow is set up initially with the worms' natural bedding or casting with the correct stocking density. The time of year will dictate the volume, the depth and the surface area you will need for the initial set-up. Because the windrow is exposed to direct sunlight and wind, it is essential to cover the bed directly with carpet or shadecloth or mulch to prevent excessive evaporation and drying and the erosion of contents.

Plastic inhibits vertical drainage, thus the contents of a windrow sitting on plastic may quickly become stagnant and anaerobic. Weed matting or a heavy shadecloth may be more appropriate from a drainage point of view.

Site location is very important for run-off, particularly with heavy downpours of rain, and it may be necessary to implement some sort of earthworks and design your site to suit wet weather accessibility. Feeding and aeration are still most important, even when it rains for weeks. If you use a tractor or bobcat for these tasks, you can end up with the beds literally surrounded by a quagmire that seems to make everything impossible.

Aeration is very important and also very frustrating because there are no sides to the windrow to stop the contents falling and rolling away. It is very hard to maintain a neat windrow and if possible it is practical to use a row of timber as a border on both sides, railway sleepers, for example. The timbers can be held in place with pegs and thus be easily removed or placed in another position as the windrow spreads. Aeration is the most important husbandry practice in windrow farming to counteract the compaction resulting from heavy rainfalls or any consistent watering. The contents of an unrestricted windrow are constantly spreading and settling because of the watering.

On the larger scale of windrow farming, a sprinkler system is essential simply because of the time involved in watering such a large surface area.

Regardless of how you water, the choice of covering directly on the bed is important. Think about the practical implications of removing covers to water and replacing them after watering—it does not make any sense, particularly when we are talking about large surface areas. Your choice of covering material should allow for the penetration of water and yet be heavy enough to stop the wind from blowing the covers away, to avoid the excessive use of counterweights such as bricks and timber.

When covers dry out they become temporarily impervious to water. It can be a difficult and frustrating job to re-wet them to the point where they will absorb and allow water to drain through them again. The windrow shape (coned or domed) causes water to run off very quickly and concentrate at the sides of the bed, creating constantly excessively wet areas. This boggy situation underfoot is frustrating and also encourages prolific grass and weed growth, which must constantly be whipper-snipped and/or mowed.

The type of animal manure that you use (e.g. sheep, cattle, horse) and its freshness, before being attacked by micro-organisms during composting, must be considered. These manures, which contain a fair amount of fibre, should be applied in a semi-moist, loose and friable form. To do this over such a large area the manure must first be drained and partly composted before it is broken up with a fork or rotary hoe. It is then applied in blanket form across the entire surface area of the bed to a depth of no more than 50 mm (depending on stocking density). Any food given to the worms must be eaten out within 5–7 days so the depth of food applied must be related to stocking density.

When the accessible nutrients are eaten out of the food supply you will be left with a fibre/cellulose mulch layer on top of the bed, which is better than hessian bags or carpet. See Plate 9. This sealed layer protects the eggs, slows down evaporation and protects against extremes of heat and cold, plus you can water through it.

As these manures contain a certain percentage of fibre that will not compress until eaten out, it retains oxygen and provides a working layer. The amount of fibre that will be residual in this layer will depend upon the period of composting (microbial breakdown).

Through sensible management of what is a difficult management system you can ease but not eliminate two difficult husbandry practices, i.e. aeration and covering/watering, simply by the right choice and preparation of a manure source.

It will still be necessary to aerate or toss the beds to get oxygen back in after heavy rain or simply to bring uneaten food to the surface or to encourage aerobic composting of accumulated fibre. Whenever you toss the bed it is very important to cover it again with a loose friable layer of manure to protect any exposed eggs and encourage the worms back to the surface.

We now have this beautiful feed over the top that is loose and friable, or a mulchy layer that we do not want removed by wind or eroded away by water. It is prudent to protect this layer by using, for example, 70 per cent shadecloth. This is placed directly on the bed and pegged down with tent pegs or similar every 2–3 metres along the side. This holds everything in place, can be watered through and also eliminates the bird problem. In addition, have you ever pulled back the covers on a windrow and had a snake come out at you? This heart-stopping moment is practically eliminated by the use of shadecloth.

Love them or hate them, windrows should be regarded as, and only set up as, huge surface area breeding beds.

Stocking density

The stocking density for windrows should be considerably less than 2 kg per m^2 (6 oz per ft^2) because of the 'walkout factor'.

Once the eggs begin to hatch, it is important to slow down feeding and begin baiting so that the adolescents can be placed into separate growing beds. Because the baiting will also attract the mature worms (if you do not use weaning trays), you either use a harvester to get the majority of adolescents out or write off a certain percentage of mature worms. Place the lot in your grow-out bed where you can supplementary feed to get the best results. Constant thinning of the windrow is necessary to reduce the stocking density.

As inefficient as windrows are, if you have access to a lot of space, on a site with good drainage and good run-off, where you can set up beds within reasonable distance of each other, thus allowing you to capitalise on the walkout phenomenon as the worms move from bed to bed, then the windrow system can work for you with very little capital outlay, even without the use of grow-out beds.

Windrows can also be a useful adjunct to a standard bedding and management system, especially if you use them to coincide with the spring and autumn breeding flushes. Simply lay plastic, shadecloth or weed mat down with the minimum amount of bedding and the correct stocking density for approximately six weeks at the beginning of the season, when it is apparent that the 'flush' has begun. This simply gives you an extra surface area to play with while times are good.

At the end of the cycle, physically pick the entire contents up and place in a spare bed. The returns from this short programme far exceed the losses due to the aggravation of walkouts and patches of dead grass where the windrows have been.

Raised beds

Growing beds

The only differences here to a normal bed sitting on the ground are the advantages of good drainage and easy collection of bed run-off, no problem with roots, good air circulation under the bed, while the bed should more readily remain aerobic. See Plate 2.

The disadvantages are the cost of making a structure strong enough to hold a lot of weight; the possibility of worms escaping from underneath the bed; and height being restricted by the physical constraints of working the bed.

Breeding beds

The advantages of raised breeding beds include the fact that the bed can be elevated to a comfortable working height, i.e. waist height, because of its depth (approximately 150 mm or 6 in). They also have good drainage, they readily remain aerobic and there is easy collection of bed run-off. See Plate 4.

Raised breeding beds are ideal for hotter weather with air circulation over the top and underneath creating the 'Coolgardie fridge' or water-bag cooling effect—good for breeding.

The disadvantages are that you require a strong structure, the beds may be too cold in the cooler months for breeding because of the increased air flow, and worms may escape.

21-day cycle—the tray system

This is simply a micro breeding bed system that consists of many trays that are stackable with a lot of potential surface area (this is what breeding is all about). Even though this is a micro system, it must still meet the basic environmental requirements that satisfy Nature's Equation, i.e. surface area, volume, stocking density, pH, moisture, aeration, drainage, food, temperature, humidity, etc.

A tray 300 mm x 500 mm (12–20 in) should take a stocking density in the vicinity of 200–300 worms. If you have 100 of these trays, that is the equivalent of 15 m^2, or 3 lots of 5 m^2, breeding beds. The trays can be of any size—just so long as they are manageable. See Plates 10 and page 35.

The trays can be set up where there is convenient cover and moved from location to location to suit seasonal temperature variations.

The cycle works on the basis that if breeding occurs on the first day, hatching (under ideal conditions) will occur on the 21st day. This is why it is called the 21-day cycle. On the 21st day you place the contents of the boxes into an incubation/grow-out bed (after removing all worms) and set the trays up again for the next cycle.

Incubation bed

If you operate a rotational system of management (see Chapter 4) the incubation bed is your most valuable bed—the success of your worm farming venture depends entirely upon the eggs hatching with maximum viability.

Under ideal conditions, an egg can hatch after 21 days with the maximum amount of hatchlings (subject to health, condition, and fertility of your worms). After this period of time, the number of hatchlings or the viability of the egg will begin to diminish, to the stage where only one worm will emerge from the egg.

It is important that the eggs (e.g. from your 21-day cycle system) are collectively deposited into a bed where they can incubate and hatch in the shortest period of time as possible.

Incubation requires a constant temperature and moisture level (humidity) and time. Bedding temperature should feel lukewarm to warm to the touch.

For better control, incubation beds can be located indoors, protected from the elements, warmed by electric cables or thermostatically controlled electric blankets. Once the eggs have hatched, the heat can be turned off and the husbandry practices that favour grow-out conditions can be implemented.

If you have access to only one or two incubator beds with cables or blankets, you need to be able to bring the hatchlings to the surface for easy extraction so you can re-use the beds for the next cycle.

Incubation beds without artificial heating can simply be given a name change and become grower beds without moving stock.

Polystyrene breeding boxes An economical and practical way of establishing a 21-day intensive egg production breeding cycle. Roy has used polystyrene broccoli boxes, stacked in rows and layers and separated by air space and timber slats. The stack slopes toward the front and each box has a hole in its base to allow for drainage in the event of over-watering. The rectangular section cut out of each lid is replaced with a piece of flyscreen mesh or fine shadecloth taped to the under-surface to provide ventilation and prevent escapes. (Courtesy Roy Ruckman)

The bonus bed

From a practical point of view, every grower should have a bonus bed. From a management point of view, it is probably just one more bed that you simply do not need.

Nevertheless, you always need somewhere to put a handful of worms not of the right size, a bucket of rubbish with worms in it, deteriorated coverings (bags or underfelt) mixed up with a few worms, or mismatched material, and this somewhere usually ends up being your bonus bed. It is placed somewhere out of the way but within striking distance and is usually neglected simply because its contents are regarded as acceptable losses. This bed will end up surprising you and the sale of the extra worms makes a nice bonus.

The bag bed

Hessian or hemp bags are marvellous as an internal covering for the beds, but they deteriorate quickly and need to be replaced every 3–5 weeks. The remnants can be placed in a special bag bed.

The bag bed will undoubtedly become the best breeding bed you ever have. The fibrous hessian is naturally interwoven with air-holes and tunnels that allow the worms free movement. The bed has a layered structure with varying levels of temperature and insulation that also retains and drains moisture. It is the ideal artificial base environment for the laying and incubation of eggs.

As brilliant as this situation is, it is also totally unmanageable, and should be viewed as a separate entity to the rest of your manageable farm—it is another bonus bed.

The sole purpose of the bag bed is the breakdown of bag remnants; its surface area and stocking density should match this requirement. To keep things going you will occasionally have to sprinkle on a shovelful of high-protein manure as a food supplement for the worms and to encourage the micro-organism population (remember that a nitrogen component is essential in the composting of this fibrous material).

CHAPTER 4

Breeding Management Systems

An efficient management system is the key to any business.

Worm husbandry is livestock husbandry, where big is not necessarily better.

Just because you can't see the worms doesn't mean that everything is okay. Because you can't see them means your husbandry practices have to be even better than the husbandry of livestock that you can see.

You need a rotational system of management, moving the worms from bed to bed, just like moving sheep or cattle from one paddock to another and back again.

- A stale bed is of no value to your stock or to you.
- A bed to pit run worms is no good for breeding, growing or harvesting. (A 'pit-run' bed contains worms of all sizes.)
- A bed that is full and cannot be fed or worked or that impairs drainage is no good.
- A growing bed with little stock in it is no good.
- A breeding bed that is not producing is no good.
- A bed sitting idle is no good.

Management must start from Day 1. Every business must have a business plan (whether it is on paper or not). This plan must include a marketing strategy, which is the aim of your breeding/growing programme, but must also include an operational or management plan covering how many beds are required, which ones are to be breeding beds and which are to be growing beds; what rotational system is going to be adopted to get the maximum number of harvestable worms on a regular basis; and so on.

System 1—The multi-functional bed

This is the simplest system, where the one bed is multi-functional, set up with breeding worms which stay there until the bed is literally full of bedding. In the meantime, the worms lay their eggs, hatching occurs on an ongoing basis and somewhere along the line you harvest whatever you can. The cycle goes on in this fashion until the bed is full, when you simply

SYSTEM 1

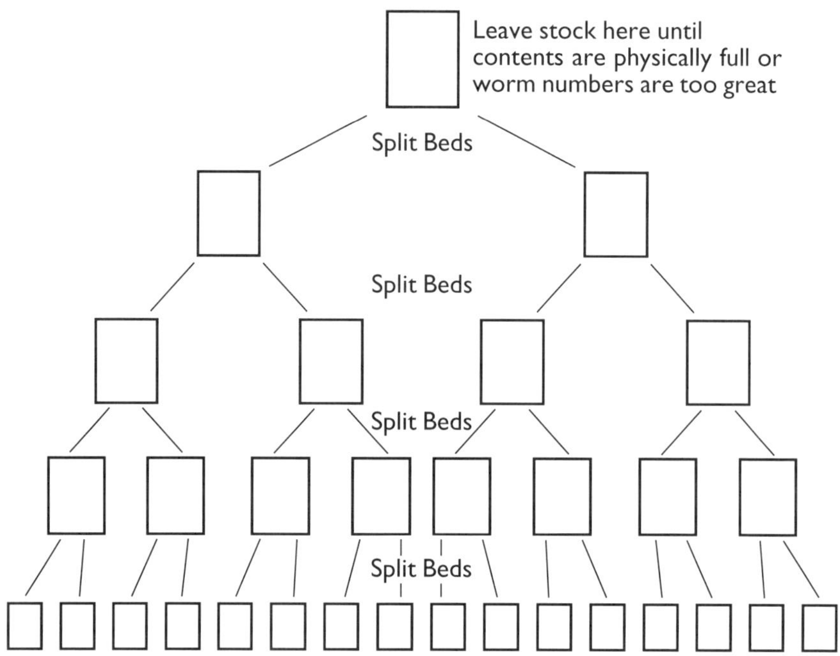

(Silly isn't it)—You end up with more than you can poke a stick at.

split the bed to reduce the volume. Now instead of one bed, you have two, and so on.

There is no management plan involved here and you come out the loser because you are harvesting worms of all sizes, including breeders and adolescents; you could end up supplying 6000–10 000 worms per kilogram instead of the 2000–2500 per kilogram that you should be aiming for.

You will have initial breeding success because you have probably started with the correct stocking density and egg production will be realistic. Once these eggs hatch the stocking density will increase dramatically after quite a short period—this is the danger period, when breeding will slow down and practically stop. You will be under the illusion that you have a lot of worms (which is true), but because the breeding has stopped you will have no ongoing supply for the months down the track. Only when you have completed a harvest and reduced the stocking density will the mature worms begin to breed again.

System 2—Weaning (removal of adolescents)

It does not make any sense to leave adolescents in with breeder worms. They are not dependent upon their parents for survival. Once they hatch out of the egg, they are on their own. They have different feeding and social requirements to the mature breeder. An adolescent is a sliver of protein with a voracious appetite which can grow at an astonishing rate when conditions are right, i.e. mass numbers and concentrated foods.

You can still operate a multi-functional bed where the breeding worms stay forever but you must extract the adolescents on a continuous basis by using a weaning fabric.

SYSTEM 2

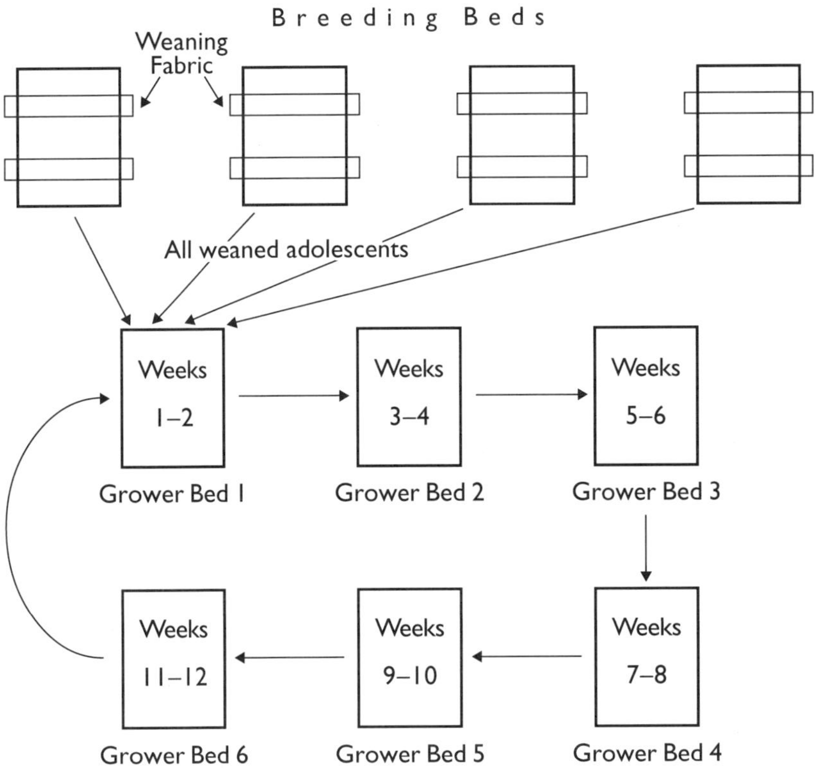

All adolescents stay in their assigned beds until marketable size whereupon the bed is cleaned out and continues the loop.

Weaning fabric stays in place for 2–3 days for maximum effect before depositing its contents into the appropriate, grower bed, then being set up and placed in a new position on the same breeder bed with fresh bait.

Weaning fabric As long as the beds are not overfed and the worms operate in the top 150 mm (6 in), weaning is a simple procedure of covering alternating areas of the bed with a fine aperture synthetic cloth (90 per cent shadecloth or a similar product) which is baited on top with a loose, friable and desirable food. Only small worms can fit through the fabric to get to this food, which should be covered with hessian or something similar to prevent it drying out. The captured adolescents (and their food) can then be placed in a separate growing bed.

The number of growing beds you require will depend on the diversity of worm sizes coming through the fabric (which in turn depends on the size of the apertures in the fabric), and your marketing/harvesting requirement (once a week or once a month, for example). A growing bed containing worms with too much variation in size means the period of intensive feeding required to produce worms of consistent marketable size will be too long. I suggest that every two weeks you put adolescents into a new bed so the sizing of the worms is theoretically two weeks apart.

It is well and good to use this system of extraction, but if no breeding is taking place, you will have no eggs to hatch; if you have eggs but the incubation temperature is not right, you will have no hatchings. There can be, and will be, periods where only very small growing beds are required to handle the quantity of adolescents coming through.

In effect, what you are doing under this system is trying to maintain breeding stock density at the right level, not extracting them and selling them off. Theoretically, you are extracting all the adolescents (the marketable worms) so there are no eggs in the grow-out beds to dilute sizes. In reality, however, you cannot get all the adolescents out and they grow into breeders. It is advisable to regularly test-weigh the bed for stocking density and adjust back to optimum levels.

System 3—Combining breeding beds

This system employs a 'no frills' approach to management which eliminates weaning and harvesting—there are inefficiencies in the system but it certainly has its advantages.

This system starts with every bed being a breeding bed. You set each bed up with the minimum amount of bedding to suit the season, and the right stocking density. The worms breed and stay in there until the eggs begin to hatch (about one month). At this time you must make a judgment as to whether or not the bed is 'burned out' (breeding has slowed down) or is still producing. If it is still breeding, leave it alone for a while to get extra results without adding any more food. It will burn out eventually, so either monthly or at your discretion, physically combine all breeding beds into

SYSTEM 3

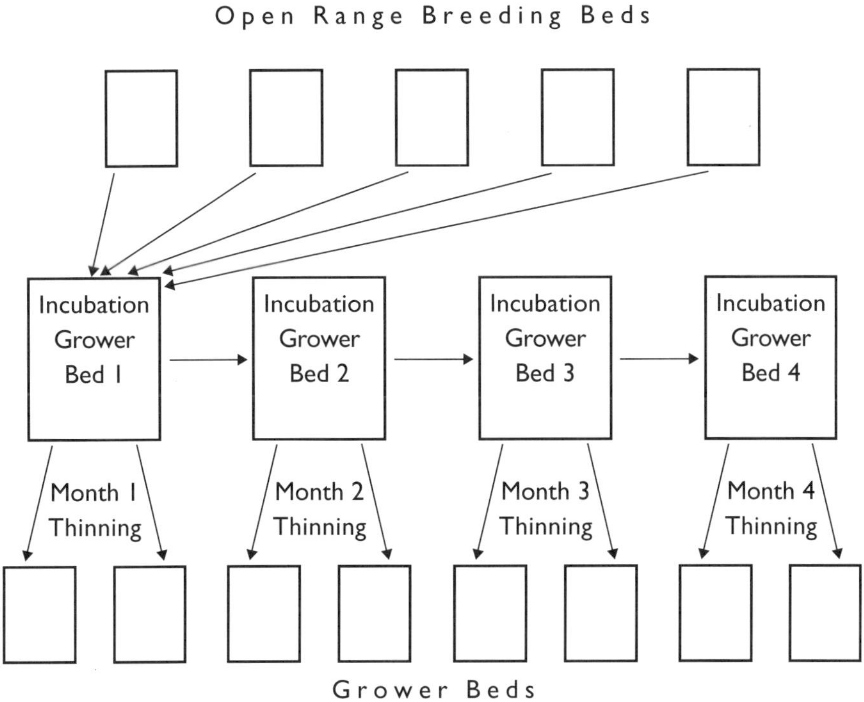

Empty entire contents of all breeding beds into incubation grower bed No1 and then reset up breeding beds with fresh worms and bedding for the next cycle that will ulitimately be dumped in bed No. 2.

Thin incubation beds to suite marketable stock management of 10 kg/m^2

one large hatching/grow-out bed. This bed (or beds) must be of sufficient size to take the contents of all the breeding beds on this cycle and allow sufficient space for the grow-out period to harvest. It is essential to use the highest quality primary food sources and supplementary foods.

In this system you 'write off' the breeders as part of a convenience cost—no weaning or harvesting required. They may continue to breed for a short period—hatching will occur and the stocking density will increase to the point where breeding stops. The temperature of the incubation/grow-out bed must be increased to lukewarm straightaway and maintained at that temperature to favour incubation/hatching and stop the adult worms from breeding further.

The scene is now set for massive numbers of hatchlings and overcrowding. As they begin to grow, the worms will need to be thinned out to the 10 kg per m^2 level (30 oz per ft^2) to allow them to reach their optimum genetic potential and manageability.

Now you have all these other empty beds to set up as new breeding beds to begin the next cycle that will again rotate itself.

It is important that you set up the new breeding beds with fresh stock at the right stocking density, so it is advantageous to have a small bed of primed potential breeders on hand.

System 4—Rotational management

This is an extension of the combined breeding bed system but takes management to the next level of operation so that all wastage (with the exception of acceptable losses) is eliminated. It requires a set programme for the harvest of worms at regular intervals and the setting up of beds to assure the supply.

All beds are set up initially as breeding beds and timing is staggered. Working on the one-month cycle, allowing one week to settle in and three weeks to lay, set up two or three beds per week. Run all the beds on the same cycle through a harvester (or light-extraction) to capture all (well, 99 per cent) of the breeding worms. The bedding with the eggs is then placed into an incubation/grower bed with the temperature maintained at lukewarm to induce hatching.

Once the breeding beds are emptied, they can be set up again immediately using fresh bedding and the same worms (if you do not want to rest them). Allow approximately one week for them to adjust to their new bedding, enter into a rising plane of nutrition and reach point of lay.

It is inefficient to have a breeding programme and not be able to quantify what happens during the cycle of egg production, incubation and hatching, or what happens in the growing cycle in regard to weight. In this system, we end up with three separate entities:

- Breeding beds—quantify with an egg count
- Incubation beds—quantify with a hatching count
- Growing beds—quantify with a stocking density check weigh.

If you cannot get your marketable quota of stock, it is important to know at what level of operation the problem lies so that you can take steps to rectify it.

In the incubation bed (assuming three weeks under ideal conditions to hatching), in theory there will be a three-week differential in size; in practice, allow one month for final hatching to occur. This is followed by 8–12 weeks of growth to achieve marketable size. The incubation bed will not be emptied (assuming it does not physically fill due to feeding before harvest) until

SYSTEM 4

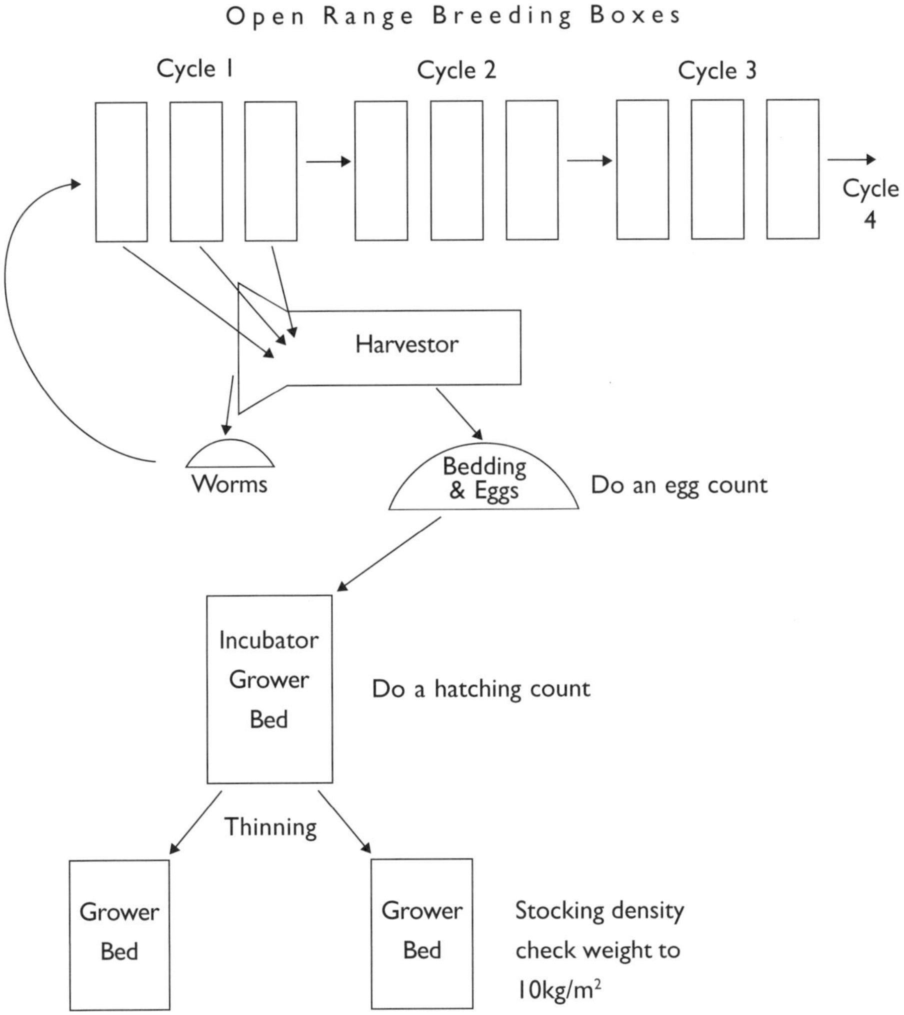

Empty all contents on the one cycle e.g. Cycle 1 into a harvestor to extract breeding worms (which return immediately to the same beds with new bedding) and then dump the sieved vermicompost with its eggs in an incubation bed for hatching to occur.

The next week repeat this cycle using contents of Cycle 2 to set up another incubator/growing bed

Once harvesting begins you will have a complete rotational system in operation

SYSTEM 5

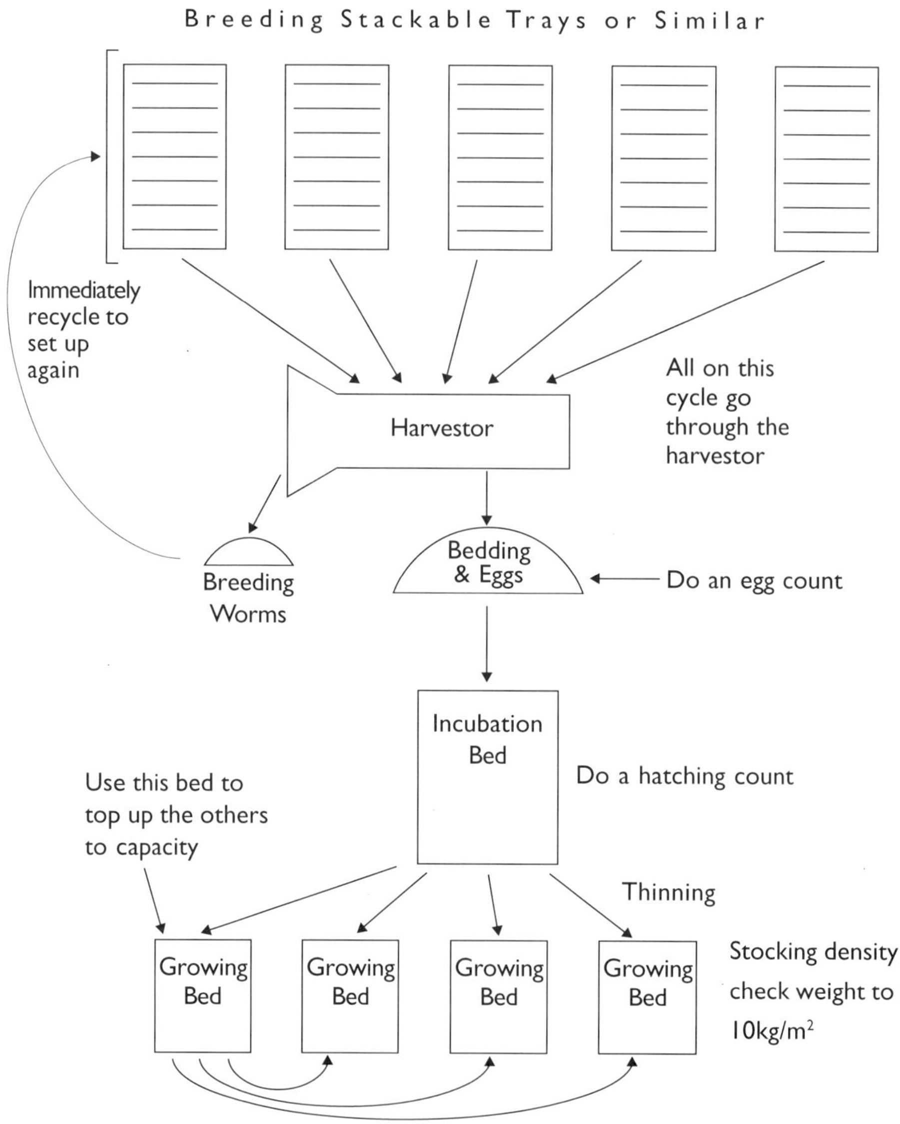

Quantifying eg

Egg count = 100 000
Hatching count = Average 4 worms / egg
Potential nos. = 400 000 hatchlings
Market aim = 2000 worms/kg
Potential kgs = 200 kg worms
No. of grower beds required = 4 lots of $5m^2$ beds carrying 50 kg worms / bed for this 21-day cycle.

harvest 3–4 months after setting up for incubation. Once emptied the bed can be set up again as a breeder or an incubation bed to resume the cycle. If the bed does fill before the anticipated harvesting date it is important to have a spare bed to which you can remove the working layer and 99 per cent of the worms. The emptied bed is thus freed up for another cycle.

The number of beds you will need to achieve your 'market quota' under this rotational system cannot really be worked out until you do your first egg count—and this will totally depend upon the seasons and your management expertise.

System 5—21-day cycle rotational management

This system, an extension of the rotational system of management, is a far more intensive method of production. It eliminates all the breeding beds required in the open range systems.

The breeding beds are replaced with stackable breeding trays, many of which can be fitted into a very small area. The intensive breeding tray system can be operated within the confines of a shed so it clearly defines a breeding section which always remains the breeding section. The breeders are seeded into their trays at the correct stocking density with the prepared bedding where they stay for 21 days (assuming they lay from Day 1) before being emptied into an incubation bed (after removal of the breeding worms) to induce hatching.

Do an egg count to estimate the stock you can expect upon harvesting. This gives you a guideline for adjusting breeding tray numbers to suit your marketing requirements.

Once the eggs have hatched, husbandry practices are changed to suit the grow-out conditions, the incubation bed now becoming the grower bed.

The only beds required for this kind of operation are incubation/grow-out beds; their size and number can be tailored to suit your marketing time-frame and the expected quantity of worms coming through with each cycle.

CHAPTER 5

Husbandry Practices

A worm does not have a thinking brain. It is a creature that survives on primal instincts and is very easily stooged. A worm does not know what time of the day it is or what time of the year it is, or what season we are in. It reacts to the environment of its bedding and can be tricked into believing that every day is spring, the time when Nature's Equation rules. The worm's two basic instincts are to eat and to reproduce. Both of these it does very well. The idea of farming worms is to balance these two elements for maximum health and breeding to occur.

Worms are livestock and raising livestock is all about managing them efficiently. It is no good having, for argument's sake, five hectares of manure and worms if you can't manage them. Good management begins from Day 1 with the introduction of your stock and must be maintained through to harvesting (which should be quick and easy). If you haven't set up the system properly, harvesting will be a nightmare and you will get no economic return from your investment.

There is no sense having something you can't manage.

Stocking density

This means the amount of stock to a specific area. Because compost worms are surface feeders, we can talk about square metres rather than cubic metres.

Breeding beds

In breeding beds only a bare minimum stocking density is required, 1–2 kg worms per m^2 (3–6 oz per ft^2), depending upon bed volume and time of year. In spring, approximately 1 kg per m^2 or less. Worms have particular space requirements and for the most efficient use of a bed you must ensure that they use the entire surface area—tossing the contents of the bed on a regular basis ensures this distribution.

In large beds, worms naturally congregate to the sides, which can leave the major part of the bed unstocked. You end up with too many worms in one location, the signal for overstocking goes out and they slow down their breeding. Minimum stocking density is thus critical to breeding.

Growing beds

In your growing or fattening beds it is essential to have what would appear at first sight to be an extraordinary number of worms, the opposite of stocking density in breeding beds. Again, stocking density is related to the time of year. A comfortable density is 10 kg per m^2 (30 oz per ft^2); this can be higher subject to other factors. Worms thrive and perform very well in large numbers and will reach their optimum genetic potential of 2000–2500 worms per kilogram in these large numbers.

Worms in these numbers require a lot of good quality food, supplementary protein food, water and good drainage, and oxygen.

A high stocking density facilitates easy management, allows optimum use of space, and makes for easy harvesting.

Dangers of high stocking density

Worms in these high numbers generate a lot of heat, so be careful. If the beds are too heavily stocked the worms will mass just below the surface like a blanket across the whole bed. They will become too hot and the bedding under them will become quite warm. Smell them—if they are in survival mode this is a warning to act quickly (see page 96 and 97).

This is one of the reasons why it is important to continually toss the beds and relocate the worms over the entire surface area, as otherwise it takes a long time for the massed worms to work their way back to the middle of the bed. A lot of losses can be sustained through overheating in a mass of raw worms.

Once the worms are in 'survival mode', particularly when jammed into a corner, their exits for escape are limited and this makes the situation worse. Worms have no common sense, nor do they use logic. Yarded sheep being worked by dogs exhibit fear and push against one another, resulting in the ones at the back, particularly in the corners, going down and being trampled on and suffocated. If you take the dog out of the equation, it still takes forever for the flock to disperse and you have still more sheep going down and dying. To rectify the situation you must physically disperse the flock. It is the same with worms—unless physically dispersed they can die.

There is no leeway for marginal management with a high stocking density. The blanket survival mode can present itself because of insufficient oxygen in the bedding, overwatering, overstocking (due to natural growth rates), overheating internally, general bedding compaction, or because of bad feeding practices that result in uneaten food souring and composting in the bed.

Stocking density may most easily be related to sheep and cattle in a paddock. How many sheep or cattle can you put on a one-hectare block? What is the block's carrying capacity? The paddock must carry the animals for one year to complete the cycle of seasons without (hopefully) supplementary feeding.

A number of variables come into play: rainfall, soil type, vegetation cover and variation of species, availability of water, summer and winter temperatures and geographic location. All these factors together determine the amount of feed that will be available at any point of time. A paddock that is lush and green in spring could carry, say, 100 head of sheep for 2 weeks; in winter it might only carry, say, 5 head of sheep but year round might carry 10 head overall.

A worm bed is no different—we are still talking about stock, a defined area and food supply.

We can increase the stocking density of that one-hectare paddock with 10 head of sheep-carrying capacity by supplementary feeding in the lean times, irrigating, pasture improvement, strip feeding, and fertilising.

Likewise, the management of worm beds can be adjusted to achieve an improvement in performance, including stocking density.

Stocking density check weigh

This is a simple procedure that will give you an approximation within 10 per cent of the stocking density of a bed. Beds which are managed correctly will have an even distribution of worms across the entire surface area. If the centre is too hot the worms will be at the sides but they will still be evenly distributed (with the exception of the corners). Mark a section of half a square metre (or whatever area is appropriate) which goes right across the bed, and remove the working layer, the top 100 mm (4 in), which should include 99 per cent of all worms in that area. Remove the worms by light-extraction and multiply their weight by the size of the bed in square metres, remembering to allow for the area you have extracted. For example, if you have a five square metre bed and your took a sample over one square metre, multiply the result of the sample in kilograms by five; if you took a half square metre section, multiply the result in kilograms by ten.

It is very important to know the stocking density in breeding beds so that you can adjust the stocking density back to where you want it by extracting surplus stock.

It is also important to know the stocking density in growing beds to give you an estimation of the quantity of worms coming up to harvest. If you know that your adolescents are half-grown and your stocking density check weigh reveals 10 kg per m^2 in a 5 m^2 bed (50 kg), you know that you have

PLATE 1 Rotational layout under deciduous tree A typical back yard set-up. These above ground beds are positioned under and around the fringes of a large deciduous tree. Some of the beds have a plastic base-lining, with internal assisted drainage, as protection against invasive roots. All of these beds receive full sunlight during autumn and winter (when the leaves fall off) and a mixture of shade, filtered light and full sun during summer and spring. Bed location such as this suits the rotational system of management.

PLATE 2 Off-ground beds These above-ground beds with escape-proof lip have been paired for ease of covering. The height of the covers allows for user convenience, good air circulation and natural light while providing protection against direct sunlight and rainfall and cold southerly winds. The beds are elevated to allow ease of working (no tree root problems) with sufficient angle to allow for good drainage of the leachate. John's beds sit on two sheets of galvanised corrugated iron which are supported by bearers that sit on bricks. The corrugated iron is overlaid with a heavy duty 10 mm (0.4 in) square plastic mesh which is in turn overlaid with fibreglass flyscreen mesh. The combined strength of these two meshes prevents the contents of the bed collapsing into the grooves of the corrugated iron and allows the filtration of the leachate through its matrix. In the event of a build-up of sediment in the air space of the corrugations a slight overhang of the corrugated iron at the high end allows access to a high-pressure hose to flush the drainage system. (Courtesy John Vickery)

PLATE 3 Above-ground beds This outside, exposed, on-ground bed is 50 m long x 3.8 m wide (46 yd x 3 yd). It is plastic-lined on the sides only and has an escape-proof lip around the whole perimeter. The bedding is domed to allow for the quick run-off after heavy downpours. The gully facilitates the flow of water to the lowest point and has provisions for drainage outlets along the way (agricultural pipe with sock). The surface is covered with 70% shade-cloth, primarily to combat the aggravation caused by birds, which has the bonus effects of holding the working layer together, can be watered through, allows oxygen exchange and can be used as a harvesting tool.

PLATE 4 Raised breeding bed This raised breeding bed allows Diane to work at waist height with a minimum of back bending. The base is made of slats to allow drainage of the leachate into a receptacle at one end via an impervious plastic sheet which drapes under the bed. The A-frame with its generous height and shade-cloth cover allows for good ventilation and shading. A micro-irrigation spraying system on a timer is attached to the internal apex of the A-frame. This set-up is absolutely superb for summer breeding. (Courtesy Diane Tatum)

PLATE 5 Pollard feeding This 2.1 m x 1.2 m (7 ft x 4 ft) on-ground bed with a surface area of 2.5 square metres is a handy size and shows the sensible addition of an escape-proof lip. The seat is tightly coach-screwed to its vertical member and cushioned with a silastic compound. The internal lip is secured with galvanised nails and also cushioned with silastic. A layer of lacquer (continually maintained) has been applied to prevent seepage to the undersurface. This photo also shows the sensible method of applying pollard in dollops, with sufficient uncontaminated bedding space between the dollops. This is a high stocking-density bed containing 25 kg of worms; the generous amount of finely chopped fibre visible on the surface reflects the friability of the structure required for the interchange of gases and moisture retention.

PLATE 6 Indoor beds These four beds are each 35 m long x 1.6 m (32 yd x 1.5 yd) wide. They are plastic-lined with two 50 mm (2 in) ag pipes (with sock) running the entire length and discharging into an aerobic digester. The distance between the beds allows enough space for a wheelbarrow only. The beds are positioned within an old sawtooth poultry shed which provides both ventilation and protection against the elements. These beds do not require an escape-proof lip because electric lighting on a time clock is installed in the shed.

PLATE 7 Knotted worms Linking or knotting is the act of reproduction in Red or Tiger Worms. It entails both worms slipping their heads under their counterpart's clitellum or 'saddle'. The head protruding from one clitellum can clearly be seen. Their high gloss sheen, healthy colour and general appearance indicate that these worms are in excellent condition. (Courtesy Robert and Sally Howlett)

PLATE 8 Tiger worm with eggs A close up clearly showing the red bands that give the Tiger Worm its common name. This worm is entering a state of distress out of its moist environment—it has a sticky low lustre appearance due to exposure to light and wind. The eggs (which come in a variety of shapes from oval to round) are all similar in colour, indicating they were laid by different worms at about the same time. (Courtesy Robert and Sally Howlett)

PLATE 9 Windrows on mine-belting Adam has used a very heavy and resilient mine-belting as a bed base. The beds are approximately 20 m long and 1.8 m wide (65 ft x 6 ft). The edge of the belting is elevated with bricks to contain the bedding and facilitate the drainage of leachate. The site was first contoured, levelled and sloped to provide optimum accessibility and drainage. Each windrow has its own automatic sprinkler system operating on a timer. A fibrous layer over the surface of these windrows, resulting from Adam's manure preparation techniques (see page 32) makes other coverings unnecessary. The entire area can be floodlit by a system of strategically positioned lights operated by a timer. (Courtesy Adam Loel)

PLATE 10 Reln breeding boxes An example of an intensive 21-day cycle using a commercially available stackable tray system which allows you to maximise on minimal space. Each tray (with the exception of the bottom one, a collection tray that can be fitted with a tap) has a grated base that allows drainage through to the next level. Each tray is set up with the correct stocking density and the minimal amount of bedding/food to last 21 days. In the event of adverse conditions the worms can and will move up and down to find their right level. These trays can easily be rotated or relocated, given supplementary feed and water if necessary.

PLATE 11 Breeder boxes These breeder boxes, approximately 2100 mm long x 600 mm wide x 150 mm high (85 x 25 x 6 in), are a comfortable working height. Even when full of manure these 21-day breeding units can be easily moved by two people. The exposed nature of this system allows the use of fresh manures. The bases of the boxes are made of perforated zincalume sheet metal to facilitate drainage and oxygen absorption. Semi-fresh friable dairy cow manure to a depth of approximately 100 mm (4 in) is seeded with 2000–4000 worms per box, depending upon the time of year. The beds are constantly covered with hessian sacks to prevent evaporation and egg exposure—the system is not watered or fed for the duration of the cycle. The egg production from one of these units has to be seen to be believed. (Courtesy Robert and Sally Howlett)

PLATE 12 (above left) Worm eggs A random sample of bedding containing eggs from a breeding bed at around the 21-day mark. Such egg numbers are readily obtainable from the correct stocking density of mature point-of-lay worms seeded into bedding conditions which satisfy Nature's Equation. In this instance the bedding is semi-fresh, very moist, friable dairy cow manure. (Courtesy Robert and Sally Howlett)

PLATE 13 (above right) Grower beds These grower beds, which are positioned on weed mat, are a manageable 3 m x 1.5 m x 450 mm (10 ft x 5 ft x 1.5 ft). They are constructed of treated radiata pine, again with a simple 'hitching rail' structure to support the external covering, which is rain, wind and sun resistant. A generous arch allows adequate flow-through and ventilation. (Courtesy Roy Ruckman)

PLATE 14 Brick beds These beds are constructed of bricks and sit on a concrete slab within an open sided shed. Although their structure is permanent, they have the advantages of strength and durability. Each bed has seepage points along its perimeter and the slab is sloped to one end to assist drainage. Tarpaulins positioned around the open shed can be raised or lowered to block sun, wind and rain. The shed is equipped with electric lights to prevent the walkout phenomenon. (Courtesy Pam McKinnon)

PLATE 15 Worm harvester This practical, lightweight, portable hand-turned harvester is about as simple as they come. It displays a mesh-wrapped cylinder of generous circumference to allow an adequate feed-in mouth, surface area and air flow. The discolouration due to etching visible on the internal surface of the cone (and on the entire internal surface) should be maintained for the effective operation of the harvester. Cleaning should be as simple as spraying with a garden hose. A heavy-duty rubber mat placed under the harvester ensures easy pick-up and removal of contents. Motor-driven machines with gearbox, pulley, belts, chains, etc. should be fitted with adequate guards to prevent injury.

PLATE 16 Koppers log breeding beds These neat 2 m x 1 m breeding beds are very easy to manage. They are made with split Koppers logs and stand a very shallow 3 logs high (approximately 300 mm or 12 in). A simple 'hitching rail' frame supports the external shade-cloth covering, its generous height allows adequate flow-through ventilation. The shade-cloth covering is tensioned and held taunt by heavy-duty elastic cord. (Courtesy Roy Ruckman)

PLATE 17 Worm products Worm products, other than live worms for composting and fishing, include products such as worm manure or worm and plant food (both displayed in 15-litre plastic bags), liquid worm manure (in 2-litre containers), worm logos and even worm number plates. (Courtesy Workin' Worms)

PLATE 18 Walkout in shed This photograph was taken inside a dry shed before the lights were turned on following a winter rain-shower. A cool, crisp night with 100% humidity and no light is an ideal situation for a 'walkout'. These worms, within half an hour, would be massing over the top and dropping into the dust of the aisles to perish.

PLATE 19 Tree roots Tree roots can devastate a bed in no time flat. The culprit in this case, the deciduous tree in the background, is 5 m (16 ft) away from the bed. Within four short weeks in summer, 75% of the bed's volume was choked with a fibrous mat of rootlets. This picture also shows a simple practical method of eliminating mess while emptying a bed. A hessian sack draped over the side of a wheelbarrow and the edge of the bed allows any spills, with worms and eggs, to fall back into the bed.

potentially 100 kg of worms at point of sale. If you also know that you must place half this population into another bed to let them reach their genetic potential, you will end up with 10 kg per m^2 in each bed, easily harvestable at point of sale.

Egg count

When you place bedding and eggs in an incubation bed, you should weigh the material, filling a bucket (say 10 litre) with samples from as many different places as possible. Weigh the contents. You do not have to be exact. See Plate 12.

Take a small sample, say 10 per cent, from the bucket to use as your base. Calculate the number of eggs in your sample, multiply it by the units necessary to get the bucket quantity, and then multiply by the number of buckets needed to fill the incubation bed to estimate of the number of eggs in the new bed. You now know the potential hatch—this is most important. Document all dates and numbers with minimum and maximum production estimates of marketable worms.

If a bed's optimum potential does not meet your marketable quota then maybe you consider sacrificing the next bed cycle to join this current incubation bed.

Hatching count

This is a valuable management tool for estimating projected worm numbers emerging from a particular system or cycle, e.g. 21-day cycle. It is the follow-up to the estimated egg count; for a reasonably fair estimate, a hatching count should be conducted every couple of days during the incubation cycle.

It is reasonable to assume that consistent sized mature breeding worms which enter a particular standardised environmental system (e.g. 21-day cycle) will produce a standardised result, thus we can estimate an average from hatching samples and multiply this by the estimated egg count.

To do a hatching count, pick out, say, 10 ripe eggs, that is, eggs that are dark in colour and at the point of hatching. Gently and continually squeeze an egg between thumb and forefinger to force the release hatch open; by dragging this open end across the palm of your other hand you will easily be able to separate and count the emerging worms.

Rising plane of nutrition

This expression refers to the feeding husbandry practices for breeding stock. Worms have two basic instincts: to eat and to reproduce, both of which they do very well. The trick in the management of breeding is to jiggle these two instincts for maximum health and breeding.

Neither starving the worms nor overfeeding them, the trick is to supply bursts of food so that they are continually on a rising plane of condition (weight, vitality, and health). This is a management tool employed with breeding herds of sheep and cattle before introducing the rams or bulls. Placing the females in good pastures just before the breeding season so they are gaining weight and condition maximises fertility and ovulation—this is the value of a rising plane of nutrition.

A rising plane of nutrition in the worm is easily achievable when you start with mature worms in 'store' condition (lean). A burst of food restores weight and triggers the breeding mechanism.

The reproductive process (egg development and laying of the egg) uses a considerable amount of vitality, strength and energy; obviously this energy and weight loss must be restored to get back to the starting point. After a breeding cycle is completed, the worm must first regain its initial loss (called compensatory gain) to restore health and then marginally put on extra condition to once again trigger the breeding mechanism.

In a 21-day cycle (breeding trays) where only the minimum amount of bedding is allowed there is probably enough food for approximately 3 weeks, which gives the worms time to become 'conditioned' to their environment and provides the trigger to breed. This is why you must constantly check their condition and supplementary feed small amounts of protein manures to achieve the best results.

In a multi-functional bed it is very hard to juggle the supplementary food because the adolescents have the highest priority for it. Because you do not know which sector is gaining the benefit from the food you are forced to overfeed.

The worms should come out of the breeding cycle of 3–4 weeks in the same health and condition as they went in (assuming they entered in store condition). The only physical difference you may detect is the lack of a clitellum on the breeding worms, which could be a telltale sign of burn-out (do an egg count to be sure) or simply be the result of incorrect temperature, which throws everything in Nature's Equation out of whack and defuses the breeding trigger.

Mature worms basically need only a maintenance diet plus a little extra for 'compensatory gain' after the loss of weight incurred by breeding.

Adolescents or juvenile worms need a maintenance diet plus a growing diet. They have no sexual instinct at this stage but do have a voracious appetite and should be isolated from your breeding stock so both sectors can be fed and managed appropriately.

Acceptable losses

This terminology refers to the worm and egg losses that will be encountered with any husbandry decision. It is impossible to salvage every worm and every egg and it is important to view these losses within the perspective of the big picture. Losses are incurred at every step—during the tossing procedure, stripping castings, and so on.

Tossing beds

Hundreds of worms are unavoidably speared or torn in half in this violent procedure, but the gains of aeration far exceed the sacrifices.

Emptying a full bed of castings

If your procedure is correct you will strip 99 per cent of all worms off the top layer and miss the other 1 per cent. This means that in a 100 kg bed you will miss 1 kg. It is far better to write this 1 kg off as acceptable losses and totally empty the bed of castings in the one day than to waste time, effort and space attempting to retrieve the 1 kg—it's an impossible task anyhow.

Sale of castings

Castings will always have egg capsules in them, but unless they are present in excessive numbers (do an estimate egg count) sell the castings anyway—the aggravation of maintaining the stock pile and picking out the eggs for very little gain far exceeds the loss.

Note

Frequent walkouts or migrations are not acceptable losses—they are bad management. It is no good believing that the intelligent ones will come back.

Do not fall in love with your worms

The worm industry is a livestock industry and livestock require a lot of attention. It is easy to see how people love looking after livestock—it makes them feel good—but falling in love with them is a different kettle of fish. Worm farming is a business and should be treated as such (it is all about profit or loss). If you can make a sale, then make a sale. If a bed is not performing, get rid of it and set up afresh. If you cannot sell the worms, put them on a maintenance diet or do not feed them at all. Escapees from your beds are exactly that—acceptable losses from an occasional walkout—leave them be. When light-extracting, be expedient—if this entails being a little mercenary for the sake of efficiency, then so be it. You cannot save every worm.

Everybody makes excuses for their livestock, e.g. they are not laying because it's too hot, too cold, pH is wrong, bad batch of manure, structuring is wrong, the wind is blowing from the wrong direction. It is a bit like the bad tradesman blaming his tools.

The simple fact of the matter is people fall in love with their worms and somehow forget the very basics—that this is a business. If the business is not performing that makes you a bad manager.

'On the boil'

This refers to stock in active prime condition, literally bubbling and alive in mass numbers in the working layer—these worms are primed for harvest.

Once they are 'on the boil' it is imperative to keep them 'boiling' by blanket feeding loose friable manures with the right balance of fibre every day (a little bit a lot more often), watering gently a few times a day (to keep the working layer very moist without destroying its crumbling structure) and tossing the working layer every one to two days to blend the manures and incorporate the oxygen. Once the worms go off the boil and lose their vitality and vigour, they disperse into the substrata. This can be likened to a party that takes time to organise. Everybody congregates and is active and alive because of the environment, the food, drink and music—once the elements are turned off, everybody disperses and goes home.

It takes a considerable amount of time (7–14 days), energy and money to set up the conditions to bring your worms to the boil and to keep them boiling but it only takes one day of mismanagement to lose this advantage.

The no-feed principle/miniaturisation/the starvation diet

A lot of money, time and effort can be spent on keeping worms in a prime condition. Just to keep your growers 'on the boil' for pending markets requires constant supplementary and primary feeding, watering and tossing, and all the time your beds are getting fuller and fuller. What happens if the market suddenly collapses—even if only for a short time? What happens if you wish to go away for a couple of months? What happens if you get no breeding for three months in summer or winter?

The no-feed principle is an excellent management alternative that entails exactly that—no feeding. (No feeding also means no breeding.)

The no-feed principle is actually a starvation situation which is gradual in its onset because the worms will continually work over their own castings (which change constantly due to the microbe population). Worms are very 'adaptogenic', to coin a word, and will 'size' themselves to their environment. They will not die but will reduce themselves down to a remarkably small proportionate size (miniaturisation).

If you do not intend to toss and aerate during the no-feed period it is important to keep the beds covered, moist but not wet, because the worms are in survival mode. It is important that they be allowed to roam the whole bed via tunnelling (this structure must not be destroyed by overwatering) which also allows oxygen through the bed. If you do intend to constantly aerate because of stocking density, more consistent heavier watering will be okay.

Miniaturisation is not a cruel practice; subject to acceptable moisture, oxygen, depth of bed and stocking density, miniaturised worms will stay healthy for an easy six months. The danger period becomes evident when they begin to lose their naturally proportioned shape and become emaciated, looking like a piece of string.

They will recover very quickly (3–4 weeks) to normal size and condition when feeding returns to normal. Their compensatory weight gain and rising plane of nutrition should also trigger off a breeding flush.

The synoptic weather chart

The synoptic weather chart is a valuable management tool that allows you to take advantage of and capitalise on foreseeable climatic conditions. A synoptic chart is easy to read once you know the basics, but it must be read every night so you can anticipate how fast a system is moving. If you follow the charts in the newspaper every day for a month or so you'll soon work out what to anticipate.

A synoptic chart shows:

- Approaching cold fronts
- Strength and direction of winds
- Low- and high-pressure systems and their intensities
- Rain depressions
- Cloud concentration and dispersal.

Rain approaching

- As a harvesting aid in cooler months, outside beds exposed to the elements can be gully-baited and covered with shadecloth in anticipation of walkout conditions
- As rain approaches is a good time to blanket feed—prior to the onset of rain you can stop watering the beds so they have drained (to allow for a possible oversupply of water from the rain), then blanket feed without watering
- Warning of possible heavy downpours gives you time to prepare covers so your beds do not flood.

Strong winds on the way due to intense high pressure system

- Outside work will be impossible—organise indoor shed work
- Feed prior to onset

- Harvesting may be impossible
- Could be advisable to take all covers off.

Cloud cover in summer

- Could be an ideal time to supplementary feed
- Good time to move stock around
- Good time to harvest
- Good days to aerate.

Every day is dictated by the weather. Whether you feel like working or not, certain things must be done regularly—like watering, feeding, aerating and harvesting. It makes a lot of sense to look into the crystal ball of the synoptic weather chart and whenever possible plan activities in harmony with the weather.

The five senses

Successful worm husbandry is dependent upon observation. It is no good assuming that because you cannot see any worms, or cannot see dead worms on the surface, that everything is okay. It is necessary to employ a combination of your five senses.

Sight

A picture tells a thousand words:

- The health and condition of your stock
- Whether your worms are breeding or not
- Whether there are any eggs and how old they are
- Exactly where your stock are located
- Whether they need feeding or not.

Touch

Your sense of touch allows you to:

- Test whether the beds are wet or dry
- Check the temperature of the beds
- Judge the structure of the bedding and check for abnormalities
- Judge whether raw worms are too hot or too cold.

Smell

This enables you to be aware of:

- The worms' level of stress
- The state of bedding and packaging materials
- The amount of raw or uneaten manures in the beds
- Unusual chemical reactions in the beds.

Hearing

Your sense of hearing can give you an indication of how wet and loose the beds are. When worms in large stocking densities are disturbed they make a noise as they push air out of their way. This noise indicates good structure and good moisture in the bed, and tells you that tossing is not required.

Taste

I reckon four out of five is not too bad.

Weeds and large surface areas

With the use of animal manure comes the problem of weed seeds and their germination in the beds. Weed seed is present in manures from grazing animals—sheep, cattle and horses. They survive the digestion process because of their size and resilience. The seeds possess a tough outer covering which is not broken down during mastication, and thus proceed unaffected through the digestive system.

In smaller set-ups, weeds will be killed as the beds are aerated—the tossing process will expose the root system or cover the foliage and the weeds will die.

In larger scale farming involving windrows or exceptionally long beds where the aeration process does not involve forking or turning, you may be forced to physically walk over the beds and whipper-snip the weeds (which will grow very quickly). For the beds to hold your weight easily you'll have to let them dry out a little. After whipper-snipping, smother the plant stubble by simply blanket feeding to a depth of 50 mm (2 in)—no photosynthesis can take place and the weeds will die. However, in spring, whatever viable seeds exist in your feeding manures will continue to germinate and shoot until the season has passed and you'll have to repeat the procedure several times.

Spring is a frustrating period for growers who use animal manures in beds situated with exposure to the sun. The problem will not exist where there is no direct sunlight—you will still have germination but due to the absence of sunlight the weeds will be very spindly, will amount to nothing and soon die.

CHAPTER 6

Feeding

Feeding, as all else, is all about sensible husbandry.

The golden rule is to match the food supply with the ability of the worms to eat it, therefore you need to control the quantity and quality of food being applied.

A clear distinction needs to be made as to the outcome required (worms only, castings only, or worms and castings) to match the system of bedding management you choose to use. If worms only are the desired final outcome, you do not want the beds filling up quickly with either castings or vermicompost because this creates another management problem. If you have a castings market, your choice and preparation of food will be different (to get the maximum amount of castings possible) and your management of the bedding system will be different.

> *Primary food* The main food supply, usually animal manure, that is readily available and economical to use.
>
> *Supplementary food* Concentrated food of a protein source that encourages rapid microbial activity.

It is immaterial which primary food source you use. The majority of growers use animal manures as a primary food, although compost worms will seemingly eat anything—it is simply a matter of finding the right method of preparing it. You may need to blend a couple of products together to make your feeding system work in the best way.

Methods of feeding

Your method of feeding is dictated by type of food and its preparation, stocking density and scale of operation. Food can be applied in dollops, in strips or as a blanket—there is no right or wrong way as long as the food supply is 'safe' and is consumed within a few days.

It does not make any sense to import manures or waste from outside your area if there is a supply in your own backyard, so to speak.

Use the method of preparation which you find the easiest and most economical (e.g. fresh, composted, hammer-milled, rotary-hoed, blended, mulched, liquid, pellets) and which allows your system to be managed most efficiently. Large surface areas can be blanket fed—the food is dropped on with a bucket by a bobcat or front-end loader, then raked evenly over a prescribed area. It can be applied with an auger or a conveyer belt with a spreader fitting attached to a mixing hopper or trailer and driven off the power take-off of a tractor.

It is sensible that your feeding method also be your baiting method. When you feed in dollops or strips you have a constant accumulation of worms in one area which, if desired, can be picked up very easily. This is why it is important to keep your worms up on the surface (the working layer); the only way you can do this is with sensible feeding. Everything below the working layer is eaten out, castings, and everything on the surface is food, only achieved by matching the food supply with the ability of the worms to eat it.

This does not mean that you starve your stock. The worms are spread over the entire surface area of the bed so a sensible feeding regime must be implemented to ensure all stock has access to the available food.

Strip feeding

Strip feeding is the safest and most sensible option. Strip feeding entails laying manure across the bed (up and over the hill) in strips 200 mm (8 in) wide and 100 mm (4 in) deep, for example, allowing enough space between strips to enable you to feed in this way another two or three times. This allows for the congregation of worms in a specific area and also allows those that do not wish to congregate to work over the unfed area. When you are confident that those strips are being 'worked', lay down another strip alongside the first and carry on in this fashion until the strips join together and the surface of the bed is smooth, whereupon it is advisable to toss the entire working layer of the bed.

Blanket feeding

This method involves spreading the manure supply evenly over the entire bed to a thickness of 25–50 mm (1–2 in). Blanket feeding should only be used with high stocking densities and with the right food preparation technique (loose and friable) and the 100 per cent assurance that your food supply is safe.

Dollop feeding

This is simply depositing small amounts of food in dollops over the surface area of the bed, allowing a reasonable distance between dollops. See Plate 5.

Supplementary feeding

At various stages of the worms' growth and development you will find it necessary to supplementary feed with pollard, chicken pellets, etc.

Dealing with manures

When you put in your development application to council for a large-scale worm farm, check that you apply for, and receive, all the necessary permits to be dealing with large quantities of materials with potentially noxious smells. The accumulation of manures and their storage poses additional management problems of space, smell, different supply sources and ages of manure, the possible contamination with wormicides, etc.

A sensible rotational solution is the use of three bays side by side with dividers approximately 500–1000 mm (20–40 in) high:

- Bay 1 is the working/feeding bay
- Bay 2 is the resting or composting bay
- Bay 3 is the new or accumulation bay for fresh manures.

These bays should have provision for drainage (directly on soil is okay) and covers to contain the smell, prevent flooding and/or excessive drying due to wind and evaporation.

Semi-trailer load of manure It is sensible management to order and store manures in bulk, to assure an ongoing supply. This is a 25 cubic metre load of saleyard manure—this manure can be used as is, in its moist state, or allowed to drain for a couple of days and then formed into a windrow to begin the composting process.

If you are composting the manure, extra space will be necessary to allow for aeration/tossing of the compost.

The size of the bays must suit the quantity of manures needed for your operation over a certain period, say 2 or 4 weeks. The design of the manure storage system needs to tie in with your feeding, composting, resting, and pick-up or delivery of management.

Stale manures

Manure that has sat idle for a long period and been allowed to dry out, and manure that has been through some sort of composting process, have both lost a lot of their value and for all intents and purposes are biologically dead. A good quality manure is dependent upon micro-organisms and insects (the next step up the ecological ladder). Stale manures must be rejuvenated to get any sort of value out of them. In the short term, it is okay to feed some stale manure directly to high stocking-density grower beds which are alive with biodiversity, as simply mixing it in overcomes the stale manure's shortcomings. This should not be done too often.

To rejuvenate stale manure:

- Soak with water to assure total absorption
- Allow to drain
- Add an activator to get the microbes proliferating, e.g. fresh manure, pollard, blood and bone, molasses, freeze-dried microbes—a combination of products may be required
- Mix and toss to blend the components and to add the oxygen necessary to begin the composting process—cover if desired
- Allow a couple of days and aerate again
- By this time there should be a considerable amount of heat in the pile (or windrow). This indicates good microbe proliferation and this source is now ready to be used.

The end product is not the same as fresh manure but is a product alive with the microbes essential for feeding.

Testing manures

Testing is not a complicated exercise, simply entailing obtaining a representative sample from the manure source, placing it in a dollop on the bed, wetting and then covering it for the night. If you lift the covers the next day and find worms either in it or nibbling at the edges, obviously this means it is okay. Leave for another day to be 100 per cent sure. If you find a ring of dead worms around the feed or they are not within cooee, then there is something wrong. The source should be treated as unacceptable and the manure isolated and composted for a further period before testing again.

Do not test by placing a handful of worms in a manure source. The worms will more than likely treat it as a hostile environment and will either ball up as a matter of survival or split up and wander over the surface, attempting to vacate the source even in broad daylight.

Worms must be allowed to enter a new source, in their own time, from their own bedding.

Pelletised or bag feeding (processed foods)

Feeding out of a bag makes everything so simple; to supply the basic requirement necessary for production is as easy as mixing and matching your products. Pelleted food gives you a huge payload once the food is reconstituted by adding moisture.

Feeding pelleted foods has these advantages:

- No weed seed
- Bags can be stacked in a shed
- Consistent quality and quantity
- Large payload in a little area
- Cost effective.

What is needed in a pelleted or bagged food?

Pelleted foods must provide a structural component, a primary food source and a supplementary food source. You will not get all three out of any one bagged food.

Structural component

A mulch base with a small percentage of inert fibre (non-compactive structure) is necessary to supply friability, oxygen and moisture retention. The pellets should contain a small amount of animal manure and a high percentage of leguminous vegetable matter so that the nitrogen content is high enough to facilitate breakdown to a good food source.

Primary food component

When you choose pellets as a primary food source the product must be 100 per cent organic with a high nitrogen content (3–5 per cent) and made from a manure base, e.g. chicken manure. These products are usually composted prior to pelletisation.

The value of pelletised products lies in the fact that only part of the goodness is removed when they are eaten. The digestive system of the worm is such that food passes through so quickly that its full value cannot be absorbed immediately, thus the worms work over and over their own castings to get all the value out of the food before they need to be fed again, so that a little goes a long way.

Supplementary component

Supplementation of grower beds is essential when you are using pelletised foods, adjusted according to the primary food source and seasonal temperature. Chicken pellets, cow pellets, even dog and cat biscuits, and other bagged products including pollard are suitable supplements.

Primary food source (manures)

Prior to starting your enterprise it is important to secure an ongoing primary food source. In the majority of cases, this primary food source will be animal manures which have been trial tested and proven in the past to be extremely successful for the breeding and growing of worms. Any organic product that is nutritionally sustainable can be used as a food; even if it is not nutritionally complete it can be blended or supplemented with other organic waste products to give the required result.

The scale of operation will probably dictate your preparation and feeding method—there is no right or wrong way of preparing food, so use whatever method is most convenient. A sensible approach, however, is important to the application or feeding of this prepared food.

Slurry (e.g. dairy manures)

Slurry is a liquid or slop feed that when applied results in the liquid being absorbed into the bed, leaving the sediment or colloidal material sitting on top of the bed. It is prepared simply by soaking manures in containers and agitating and mixing them to form a slurry.

There is usually enough moisture in slurry that, depending upon the frequency of feeding, it may not be necessary to water the beds at all, subject of course to adequate coverage and good drainage.

Preparation of food in this form is, of course, smelly and horrible, and requires suitable pouring containers or a sludge-pump for feeding. The mixing and storage of slurries attracts flies in the warmer months and therefore the containers should be covered.

Adequate protection in the form of heavy waterproof gloves must always be used to guard against the risk of infection from spillage or backspray entering open or exposed wounds.

Green to semi-green manures (e.g. cow and horse manures)

These manures either come directly from the animal or have sat in a heap for a few days and still retain their shape, colour and smell.

From a practical no-fuss point of view, these manures require no preparation and can be fed as is. However, they are more effective if they are broken up with a garden fork or rotary hoe to achieve a friable nature that lets the oxygen get in and allows aerobic micro-organisms to flourish.

The worms will find this more acceptable and can physically enter the manure instead of nibbling away at its edges (if applied as a pat or a solid lump).

Fresh is definitely best and on a small scale this is the way to go, providing you have a continuing access to fresh manures.

Fresh manures become very heavy and awkward to handle as your scale of operation escalates, particularly with stockpiling (smell and compaction). Green and semi-green manures attract flies in the warmer months and therefore should be covered.

Fresh manures require the same protection against infection as slurries.

Composting animal manures

As your scale of operation grows, it is essential to alter your preparation techniques or refine the ones that you have. Composting or partially composting is a 'Rolls Royce' alternative. Composting is a process of aeration and heat generation which denatures the physical shape, smell and features of the manures, while retaining an excellent microbial population, which allows you to manage the bulk with a great deal more ease.

The effectiveness of composting depends upon a delicate balance between moisture content, oxygenation and the carbon : nitrogen ratio. All manures fit this criterion.

The process of composting Create a windrow or pile of manure about 1 m (3 ft) high and approximately 1.5–2 m (6 ft) wide. If the manure is wet to the point of saturation, allow the pile to drain for a couple of days. (See page 21.)

Simply by creating the pile you have added oxygen to it and the process has already begun. Every two or three days turn the manure to allow more oxygen in to 'fuel the fire'. Once the process has started, the heat will become intense, which means that the moisture and oxygen contents are smack on.

If the structure of the compost does not rapidly become granulated, loose and friable, you will need to use a rotary hoe to achieve this texture.

The contents will go whitish/grey in colour and the manure will lose its rank offensive smell.

At the point where the smell changes, it is time to stop the composting process, test the material, and begin feeding the compost to the worms—this can happen within 7–10 days. To continue the composting process to its end is of no value.

The required result is moist, crumbly, full of air, light, and manageable by comparison to its previous raw form. Partially composted, packed with microbes and friable, it allows the worms to 'hit it' straight away.

Which manure to use?

Regardless of your manure source, there is a golden rule—match the food supply with the ability of the worm to eat it.

As an initial rule of thumb, calculate the quantity of manure required on the basis that 'a worm can eat its own body weight per day', e.g. 100 kg of worms can eat 100 kg of food per day. The precise amount will vary markedly depending upon your food source and its preparation, e.g. fruit and vegetables, contain a very high water content which will filter through the bed (after being mulched) and leave very little slurry material on top. These products decompose very quickly and are eaten very quickly, which means that the conversion ratio exceeds the worm's own body weight per day.

Washed sheep and cattle saleyard manures, partially composted, are converted at around the worm's body weight per day. Concentrated supplementary foods such as pollard and animal pellets, subject to their protein content, can seem to take forever to be eaten; the conversion ratio is nowhere near the worm's own body weight per day.

Your choice of manure is largely dictated by availability and accessibility. It does not make any sense to access and pay for manures out of striking distance.

Loose and friable manures, such as horse manure, are excellent for setting up breeding beds but because the horse has a single stomach, its digestion of plant matter is less complete than that of cows and sheep, which have several stomachs and produce much less undigested fibre. Horse manure is not a favourite for growing beds as the goodness is eaten out very quickly and must be constantly supplemented to allow time for the fibre to be broken down. Fibre can only be broken down via the microbial action and heat of composting. Composting in winter beds is desirable but no good under summer conditions.

The manures from grazing sheep and cattle, which have four stomachs and regurgitate their food, are more completely digested and are therefore better suited to the growing beds than the breeding beds. There is less fibre to contend with, so residual composting is drastically reduced. However, by their very nature these manures compact quite considerably when converted to castings and the growing beds must be worked more frequently.

Manures from the likes of pigs and poultry, which are fed on a high energy–high protein ration comprising grains (e.g. wheat, barley, soy beans) and processed animal products (e.g. meat meal, fish meal), should be treated with respect, as they retain a lot of active nutritional value and a lot of fibre. Although they can be used fresh in small quantities, they are best fed after partial composting. Both these manures generate incredible heat in a short period of time. They both have a high nitrogen content and go a long way as a food source, producing superb weed-free castings.

The dangers in the overuse of high nitrogen manures are the same as the dangers of the overuse of high protein supplementary foods—overheating, toxic gas poisoning, the risk of gut-shot syndrome (page 97).

Manures with strong smells (e.g. pig and poultry) that cause you to gag and your eyes to water, are producing poisonous gases (ammonia or methane) as a result of biological breakdown, can kill your worms if used direct. If in doubt, you always have the choice of composting or blending to produce an acceptable mix that will give you nutrition, fibre content and structural stability.

Manures that are not wholly and solely manure should be rejected. The wood shavings, sawdust and rice hulls often found in deep litter, poultry manure and manure from stabled horses, have no nutritional value. They contain a high carbon : nitrogen ratio and can only be broken down via composting, which will generate excessive heat in the beds.

Manure from beef cattle feed lots

Special care must be taken with manure from beef cattle feed lots, a source which should perhaps be on the bottom of your list. Although the manure produced by beef cattle is basically no different to what comes from a dairy cow, the supplements which are given to feed lot cattle, and the way the manure is harvested, makes it a difficult and risky product to use.

Dairy cows get the best of both worlds, receiving a concentrated food supplement while they are being milked, then turned out to pasture for most of the day to contentedly graze and exercise. The manure is collected from the holding lanes and yards daily when they are washed down, the colloidal materials sieved off and the dirty water drained into a settlement pond. The lanes and holding paddocks (not part of the wash-down system) are regularly scraped. The scrapings, along with the sieved colloidal material, constitute the basis of dairy manure, which is a brilliant food source.

By contrast, beef cattle in a feed lot are jammed into confined holding yards where they remain for 100–120 days. The yards are generally not cleaned out until the cattle have reached marketable weight and are transported to the abattoirs. Feed lot cattle are given 2.3 per cent salt in their rations to encourage them to drink a lot of water to keep their stomachs full. This eliminates the triggering device of the empty stomach that creates constant bellowing. A high salt content may kill your worms and will certainly drive them out of the beds—and does not help soil or plant growth.

Along with the accumulated manure in the yards is a concentration of urine, which you don't get in dairy cow manure. The water content of the urine evaporates, leaving behind the chemical salts. The end result is a lethal alkaline-concentrated chemical cocktail that has been stomped and ground by a lot of hooves and is devoid of any reasonable structure. This manure is very difficult to work with and sets like a rock when dried. It can be used but requires blending with other organic products (e.g. paper pulp) that are fibrous (i.e. structural), acidic and low in salt by nature. Only use it as a last resort.

Supplementary feeding in growing beds

With high stocking densities it is essential to supplementary feed with protein products. There is only a limited amount of space and a huge population of worms to use this space—a high concentrated food is required. The worm is a sliver of protein which in its adolescent stage requires not only a maintenance diet but a growing diet—these creatures have a voracious appetite.

Feeding normal manures to adolescents seems like a never-ending job and, before you know it, your beds are full and have to be cleared out before the worms reach marketable size and weight. The sensible use of supplementary feeds will considerably slow the rate of bed fill-up and will have a dramatic effect on the health and condition of your stock.

Large numbers of worms eat the guts or goodness out of manures very quickly, leaving behind a lot of fibre which very quickly builds up under the working layer, resulting in heat generation and possible souring or anaerobic conditions in the bottom layers.

Sensible use of supplementary foods entails alternating feeding or using them in conjunction with the primary food source so that the stock is forced to continually work over existing bedding while there is still plenty of good quality food on hand.

Supplementary feeding is also one of your baiting methods for quick extraction of virtually raw worms.

The nature of the carbon : nitrogen ratio of supplementary foods will induce fermentation, putrefaction and composting involving the micro-organisms found in the supplementary food. Whether or not the worms eat the food in a cold or hot (active) form does not matter. Micro-organisms and their by-products are an excellent food source.

Always leave enough room around the supplementary food source so that any gases produced can escape upwards and are not forced downwards into the bed. Always place feed on the surface after tossing, to let it 'bed in' and become partly submerged, in contact with the bed's moisture.

Once exposed to the air, decomposition will begin within a short period of time—the smell associated with microbial decomposition will attract the worm like a magnet to devour the decaying areas.

Covering supplementary food is generally acceptable in the cooler months but should be discouraged in the hotter months because the rate of decomposition exceeds the worms' ability to eat it. This is because the worms move downward to avoid the heat while the food sits on top going off and poisoning the surface of the bed.

In summer it is better to use 'cold' (non-fermenting) foods which have already been through the composting stage and are more stable, e.g. a commercial pelletised organic manure with a nitrogen content of 3–5 per cent. The beautiful thing about this is that the worms can work it over again and again and still get value out of consuming their own castings.

I cannot stress enough how important it is to match the food supply with the worms' ability to eat it. With putrefying and fermenting products allow a couple of days for them to 'go off' and then make sure the worms are controlling the food supply and that it is eaten in a couple of days. The concentration of worms around a dollop of food should be such that it is literally floating on raw worms. The speed at which they eat it depends upon microbial activity of the source.

Supplementary feeding of 'hot' (fermenting) foods, e.g. pollard, chicken pellets, in the cooler months can be used to advantage to warm the beds. Use sparingly, tossing small dollops into the bed. They will ferment below the surface and generate heat which, along with the cellulose content of the breeding bed, begins the composting process. This is the trick in winter—to maintain the heating process through the use of microbes.

Pollard and other fermenting foods

Pollard should be fed wet, placed in dollops on the surface of the bed (after a good tossing) with sufficient space between to allow a zone of natural bedding which is uncontaminated in case anything goes wrong. See Plate 5. In the cooler months, pollard can be covered with a protective layer of manure or hessian bags because the rate of microbial decomposition is slow and the danger of gases poisoning the bed is minimal.

Only apply enough to be eaten within, say, 5 to 7 days. Always rotate the area of use when supplementary feeding. Pollard has a protein content of 10 per cent, which is not high and therefore pretty safe to use. It is readily available and economical. Once the worms have acquired a taste for it, it can be a valuable tool in the overall management of your beds, from supplementary feeding to heating to harvesting.

Allow one dollop to accommodate 0.5–1 kg (1-2 lb) of worms for 5 days (subject to temperature); in other words, 50 kg worms require a minimum of 50 dollops (of reasonable size) so they can all get a feed.

The downside of feeding pollard is that it encourages the white worm (yeast worm), which you may or may not have in your beds, and which proliferates on flour/yeast products.

Putrefying foods

These are green products which are still alive, e.g. fruit and vegetables. An excellent source of supplementation, they must be put through a mulcher and broken down as far as possible to assist the putrefaction/fermentation process and to destroy the seeds. Because they contain a lot of moisture, you will have to watch that you do not overwater the beds while you are feeding them.

Apply these foods in strips, leaving a safety zone. Their watery nature means you will get 'soak down' occurring where fine active particles are leached down into the bed. Putrefaction in the bed below the working layer is a problem and the areas where the food was applied will have to be tossed and checked periodically for detection of heat and smell. Always rotate the area of use when supplementary feeding.

Supplementary food and casting production

The production of good quality casting is totally dependent upon good feeding husbandry techniques. Castings are 100 per cent eaten out in nature—this, of course, is impossible to achieve commercially because you would literally have to starve the worms by forcing them to eat their own castings over and over again. You can, however, achieve good quality castings while retaining the health and condition of your worms.

Instrument a policy of alternating feeding, i.e. feed till you have a 150 mm (6 in) working layer of manure and then supplementary feed with pollard for about one week (longer if necessary) until the working layer is worked out. Then feed manure again to get the next 150 mm working layer which you allow to be worked over again while supplementary feeding.

This is an excellent management technique, not only for getting good quality castings that are eaten out but also for setting up growing beds for summer, when it is essential to have a cooling layer of castings where the worms can retreat from the extremes of summer heat.

Feeding in the hotter months

Worms, like all other creatures, are comfortable within a certain temperature range. Once the surface working layer of the beds becomes too hot (due to direct sunlight or feed overheating), they automatically move away from the heat, down into their cooler casting where they will stay until the surface cools. Hence, they spend less time on the surface eating.

It is only common sense that your primary feeding requirement will be less in the hotter months—remember the golden rule of matching the food supply with the ability of the worm to eat it.

It is also very important that the amount of fibre/cellulose used in the summer is drastically cut down to reduce heat generation inside the bed as it breaks down through composting. Fermenting foods such as pollard should be eliminated and replaced with a cold food as they 'go off' very quickly in the heat and their gases can poison the bed. Never leave fermenting foods on the bed or toss them below the surface while the heat is on for the following reasons:

- While below the surface they generate a lot of localised heat
- They attract blowflies and maggots to the bed
- On the surface they dry out and go mouldy and must be thrown out
- From an economic point of view, it is a waste of money
- From a management point of view, it is a problem not required.

Summer conditions are spasmodic and it is not continually hot week after week. It is sensible to listen to projected weather conditions or watch the synoptic charts to make your own assessment of when to feed and what to feed. If your feeding management operates on a 4–5 day cycle, there is no reason why your stock should lose any condition over the hotter months.

Feeding in the colder months

Compost worms are surface feeders, but when the surface conditions are very cold, in times of frost and snow, they will be forced to retreat into the lower part of the bed where it is warmer by comparison. When conditions in the beds get very cold, the worms become extremely sluggish, to the point where they practically do not move.

Food on the surface that is as cold as the rest of the bed will have no effect in keeping your worms on the surface. It is essential to warm the bed up via composting and use covers in a sensible way to trap the heat inside and stop the radiation of the external cold from entering the bed. It is important that covers do not get wet, because air blowing over the top of a wet cover will create even colder conditions (the water-bag effect).

As winter approaches, gradually increase the amount of fibre being used in or with your primary food source, for the sole purpose of creating heat via composting.

If you compost prior to feeding, it is important to short-circuit the process so that the end product still contains a fair amount of fibre/cellulose. If the food product you are using is devoid of fibre, it is essential to add fibre to the beds and aerate it into the beds.

The fibre you use for this purpose in winter, e.g. lucerne chaff, wheaten chaff or a commercially available pelletised mulch (guaranteed weed seed free) must be used in a finely chopped form to make tossing the bed easier. (Think about tossing a bed full of soggy straw!) Once the fibre is in the bed, it is a simple case of altering the conditions to favour microbe proliferation, usually a case of adding oxygen and allowing the bed to drain and dry out a little.

Once the process of heating has begun, it is important to keep it going by feeding the microbes with molasses or other fermenting foods such as pollard. Fermenting foods are an ideal way of feeding and encouraging microbes. Using them, you can produce isolated hot spots or create an overall internal furnace—use them with discretion so that you do not end up cooking or poisoning the bed—and remember to match this usage with the stocking density.

Breeder beds set up in the cooler months should contain a high percentage of fibre, because they are not particularly deep and need to be 'locked up' with the maximum of oxygen in a system that will not compact, and which is heavily covered to prevent heat loss and prevent the outside radiant cold from entering.

CHAPTER 7

Watering

Common sense should be used at all times when watering. Your watering programme must fit into your work cycle. If you can only water once a day, then a fairly heavy watering will be required and you may need to cover the beds to conserve moisture until the next watering.

The amount of water needed is directly proportional to drainage and evaporation. Drainage is directly proportional to bedding structure and aeration. Evaporation is relative to temperature and humidity. The combination of these factors will determine your watering regime. Where beds are uncovered it may be necessary to sprinkle frequently to keep the surface moist.

The most common trap growers fall into arises a few months down the track, where bed height is such that drainage is impaired and the bottom half of the bed becomes waterlogged and turns anaerobic—this is one of the reasons why it is essential to routinely aerate the beds by turning them over with a garden fork.

Water source

Natural rainfall is the best watering that you can give your worms. Worms can be likened to plants that survive a dry spell—even with more than adequate watering, it is remarkable what a difference a natural drop of rain makes. If at all possible, you should expose the beds to natural rainfall for a sensible period of time to take advantage of this phenomenon.

The sensible use of misters and fine sprays within a defined structure (e.g. shed) is the closest artificial scenario that can be set up to duplicate natural conditions. Town, tank, dam and bore water (artesian and sub-artesian) that is tested as 'suitable for domestic stock', can be used successfully.

Summer heat and watering

The summer heat can literally be a killer to your worms and your husbandry practices need to be smack on.

Direct sunlight

Provisions for shading the beds are essential if you live in an area of long summers. Direct sunlight on the beds, particularly grower beds with some degree of depth, transfers or radiates a lot of heat into the bedding. Watering a bed in extreme heat will have an immediate cooling effect, which the worms (particularly blues) will sense. They will be confused because the cooling is to the surface, not below, and will shoot straight out of the bed to flounder on the surface in the sun and die from stress. Following the immediate cooling effect which tricks the worms, the sun magnifies the heat on the surface to a level higher than it was before you watered—this cannot happen if the bed is protected by heavy duty shadecloth, is inside a shed or under a tree. See Plates 1, 2, 3, 6 and 13.

This is why it is absolutely essential to enter summer with a degree of depth in the castings in the beds which the worms can escape to—this is their safety zone. Their natural instinct is to go down to escape heat. If you see worms on the surface of a bed in broad daylight you know there is something drastically wrong.

If the sun is off the beds it is safe to water as frequently as you see fit, keeping in mind that the degree of watering required is directly proportional to evaporation and drainage.

If you are locked into a situation of direct sunlight on the beds and extremely hot days, it is advisable not to water in the heat of the day. The heat will progressively drive the worms down to a level where they are comfortable, providing composting is not taking place in the bed, when the core temperature may become excessively high. In such a circumstance the worms will work their way to the edge of the beds where they will literally be massed on top of one another, generating even more heat and stressing badly. This is a horrible situation which if not rectified immediately will result in mass deaths. The emergency solution is to scoop the worms up and place them in a cooler bed.

It is absolutely essential to enter summer with castings comprising the majority of the worm beds, not to overfeed and to be particularly careful about the amount of fibre or cellulose used.

The ideal set-up for grower beds outside with exposure to the sun is a bed with shadecloth protection and a sprinkler/mister system.

Watering on-ground breeder beds Because the contents of your on-ground breeder beds are not more than 150–225 mm (6–9 in) high and tapered to the edges, watering shouldn't be a problem. If they are set up directly onto the ground, the soil underneath is stable and cool so retreat is not a problem. Drainage shouldn't be a problem, unless the beds are set up on clay, so plenty of water can be applied and the beds should remain constantly cool.

Raised beds Depending upon the depth, watering raised beds shouldn't be a problem because there is nothing to impair drainage. Stagnation of water flow and anaerobic conditions should not occur unless you have tin or plastic on the bottom, where water has to move through the length of the bedding strata to get to the other end. This is a less-than-ideal scenario.

Theoretically, raised beds should be set up so that water drains straight through. If you wish to collect the run-off then a set-up for diverting the water should be made under the bed. A base of steel or plastic mesh with an overlay of shadecloth or some other synthetic fibre to prevent the contents dropping out should allow for excellent drainage. In the summer time, raised beds produce the desirable effect of the water-bag principle, with air blowing over and under the bed creating evaporation which cools the contents of the bed.

Covers (heat and watering)

I am not a believer in covers on high stocking-density growing beds because they restrict the passage of oxygen, but they are essential on breeder beds because every time you toss the bedding you place eggs on the surface where they will dry out very quickly and be destroyed. From a management point of view, the use of hessian bags is beneficial, because you can water through them without having to take them off, they are not so heavy and allow oxygen exchange to take place. They do not last very long (3–5 weeks) but that's a convenience cost that can be absorbed.

In summary: Common sense should always prevail; don't create a quagmire to cool the beds; avoid watering in the heat of the day—worms will find their own level of comfort to ride out the heat. Frequent watering is more advisable than a one-off heavy soaking—this way water is constantly draining out, and as the water goes oxygen follows. When you water more frequently, you have to toss more frequently to compensate for the compaction that occurs.

Sensible feeding is important because the more you water, the more you are dissolving the food contents, resulting in the food being leached through the bed to then compost and generate a lot of heat. We want the food to stay on the top, so less food more frequently is the rule.

Worms will tolerate a considerable amount of heat but what is important is to observe their behaviour to assess if they have reached or are near to 'critical mass'; this is obvious if you have any of these danger signs:

- Worms on the surface in daylight
- Worms huddling or clumping together
- Stress smells (see page 96)
- Worms frothing

- Worms around the periphery of the bed, and particularly if they are jammed into the corners.

Winter cold and watering

It can sometimes become very confusing to tell whether the beds are wet or simply giving the illusion of being wet because the bedding feels cold—like clothes on the washing line. The worms' moisture requirements must still be maintained regardless of the climatic conditions.

Even though the castings retain an excessive amount of moisture, it is still necessary to have leachate draining out of the bed. This 'flushing of the system' is essential as it drains dissolved gases and toxins from the beds; as water flows, air will follow, and we need to know that 'the drain is not blocked'.

As in summer, watering requirements will be influenced by bed temperature, stocking density, bed volume, use of internal covers, etc. and will vary from bed to bed, but as a general rule watering time and frequency will be drastically reduced simply because there will be practically no evaporation.

CHAPTER 8

Harvesting, Baiting, Packaging & Transportation

Harvesting

To enable you to weigh your harvested stock prior to packing and shipping, you need to reduce the worm mass to 'raw form'. This need not be a complicated or difficult job if you manage your beds competently. There are two options, baiting with light-extraction, and mechanical harvesting.

Baiting with light-extraction

This entails placing food on the growing bed in various locations to induce the worms to congregate en masse—effective attraction may take a couple of days. With a fork you dig under the mass and place it in a bucket for transport to a table where the contents are tipped out in a pyramid shape or a long continuous row.

The worms, because they are light-sensitive (photophobic), will move to the bottom of the pile. By continually scraping and removing the bedding, layer after layer, you will eventually end up with a clump of worms free of bedding which are then weighed and placed in packaging material.

The baiting system you use here is totally dependent upon stocking density and stock placement within the bed. The choices available include dollop, strip, blanket or gully baiting; the effectiveness of each one will depend primarily upon how much food already exists within the bed and on the choice of baiting product.

If your growing beds are managed correctly, the worms are constantly in the working layer and you are matching the food supply with the ability of the worms to eat it, then you will not have a problem with using your primary food source as your bait. If there is too much primary food source through the beds and the worms are all over the place, you may have to find a more attractive bait to coax them to the surface, e.g. rockmelon, pollard, pumpkin.

The results achievable and time required to harvest a bed depend totally on your stocking density and the effectiveness of your bait. You should be able to get out, light-extract and pack 10 kg (22 lb) of worms in one hour.

Harvesting is the end product of your labour, so the initial set-up of your enterprise must be designed with this in mind. The growing/ harvesting beds should have a stocking density of up to 10 kg per m^2 or 30 oz per ft^2 (subject to time of year).

Mechanical harvesting

This section refers to the traditional trommel harvester with a cone attached. As efficient as these harvesters are, you may still need to do a little light-extraction afterwards. Consistency and moisture of bedding and an even constant speed of rotation are important for the good operation of these harvesters. Harvesters enable you to move a lot of material in a short period of time, by comparison with the traditional light-extraction method. You have the added advantages of screened castings, the weaning of adolescents and harvested worms of a constant size.

Ideally, the sieve section of the harvester is constructed of woven wire with an optimum aperture of 7 mm (0.3 in). The woven wire facilitates air circulation and its undulating surface assists the break up of bedding without interfering with flow of material along the cylinder. See Plate 15.

The process involves feeding the machine at the high end. The contents of the bed (which should be loose and friable) should break up and fall through the apertures along with eggs, hatchlings and small adolescents. The larger marketable worms will roll over and over on the fixed plane of the cylinder and exit at the cone end. Because worms are moist and sticky, they will rise up the dry cone to a drop-off point which is different to the gravitational drop-off point for any balled-up bedding which has not fallen through the sieve. If the bedding is too wet, the apertures will clog, you will still have separation, but all the bedding will ball up and be discharged out the other end with the worms.

For the most effective operation, the cone should be kept dry and therefore cannot be used in wet weather or with 100 per cent humidity. With a wet internal surface, the worms will cling to the cone and not have a definite drop-off point. The ideal harvesting situation is a dry, warm, sunny day with a gentle breeze, with the machine set up under the shade of a tree.

The cleaning of the machine should be as simple as spraying it with water, as it is necessary to leave an etched film on the internal working surfaces to prevent the worms clinging.

Common sense tips for harvesting

- Light-extraction takes time so it is important to do it as quickly as possible, particularly when the heat is on

- Worms stress once they are out of their natural environment and begin to froth and lose body weight
- Avoid harvesting in direct sunlight and wind. Although these have the desired effect in the early stages of extraction, as the bedding becomes scarce and the worms mass together they begin to froth excessively
- Harvest under the shade of a tree or in a shed with good fluorescent lighting
- High voltage lights, which are bright and hot, have the same effect as direct sunlight.

The horribles of harvesting

- Harvesting should not only be an easy task but should also be enjoyable and therapeutic.
- It is absolutely vital that the working layer be crumbly or loose and friable with a high stocking density. This allows worms to move freely and quickly away from the light if you are working on the table, or ensures that the bedding falls easily through your harvester
- The presence of grass clippings, matted fibrous material, putrefying waste, straw, leaves, garden waste, etc. makes the job of harvesting not only extremely lengthy but also truly horrible and frustrating.

Management of growing beds before harvest

Management of stocking density in the lead-up to harvesting is very important to ensure a pleasurable quick job. Your feeding techniques should be such that the worms are always in the top 150 mm (6 in) or less of the bed and everything below this is casting. You will observe that the casting (although moist) is quite compact, due to the fact that these worms are efficient little vibrators when thick in number. In very moist beds there will be no worm tunnels below this 'vibration layer' and no oxygen—the trick is to keep the worms out of here.

Prior to light-extraction harvesting, the surface structure should have a granulated appearance and be loose, friable and moist.

Prior to mechanical harvesting, it is important to bait the entire surface of the bed with a loose non-adhesive mulch product, high in cellulose but also a food source, that will break up and separate while in the machine even though it is mixed with moist bedding material. The downside of the traditional trammel harvester is the clogging of the apertures, which can be largely overcome by using such a product.

When you strip the bed it is important that you do the entire bed, not just a section, and thus get 99 per cent of the worms. If, for argument's sake, your harvest takes only 40 kg out of a 50 kg bed, leaving 10 kg behind, it will be practically impossible to get the remaining worms. You have lost the vibratory advantage of density and the worms will be tunnelling at all levels through the bed. The best solution here is to put the leftover 10 kg into another bed to increase its stocking density.

The 1 per cent of worms that you leave behind at harvest should be regarded as acceptable losses. All bedding and/or casting should be removed from the harvested bed and the bed set up again.

Alternative harvest-baiting techniques

Shadecloth harvesting

Shadecloth harvesting of surface worms is something you can do as a one-off to capitalise on wet nights with beds outside and exposed to the rain. It works very well in the cooler months. Because this method depends on the presence of rain and humidity to do the harvesting (in raw form), shadecloth harvesting probably will not coincide with your marketing dates. It does, however, allow you to thin the beds and concentrate the surface worms into a holding or fattening bed ready for an easy harvest. See Plate 3.

This wet-weather event should be coupled with gully-baiting (see below) to achieve maximum results. It is essential to have 70 per cent (or slightly less) shadecloth laid on the surface of the bed before the rain comes—which must be a decent steady rainfall followed by coolness and lingering humidity. As oxygen is depleted from the bed by the rain, the worms come to the surface through the shadecloth and wander around in an attempted walkout (an environment of 100 per cent humidity where conditions are dark, moist and cool).

Shadecloth harvesting is a night-time job done with a torch, and wearing wet weather gear and gumboots. It is as simple as folding and rolling the shadecloth to force the worms to slide to a point where they can be picked up—the synthetic non-adherent nature of the shadecloth makes this an easy task. The number of worms that you harvest is a function of the stocking density of the beds and the effectiveness of the night; this form of harvesting can be done hour after hour.

The beauty of the 70 per cent shadecloth is that when daylight comes the worms can go back through it and be protected against the elements. Your choice of shadecloth is important as some cloths have very fine strands or filaments in which a larger worm with clitellum can become snared. You can end up with a lot of damaged stock.

Gully-baiting

This method of baiting is simply the best way to collect the worms, on four counts. It takes account of worm behaviour, weather conditions, bed conditions, and is easy to manage.

The worms' natural instinct is to wander and mass together (survival instinct). If conditions are not right in the bed (e.g. too hot) they move away and accumulate at the sides or the corners, places where they can go no further. If conditions encourage a walkout they remain massed along the sides of the bed if they cannot get out over the top. This movement to the sides of the beds is an ongoing behavioural trait and can be capitalised on by gully-baiting.

Around the inner edge of the bed dig a furrow or gully approximately 150–200 mm (6–8 in) wide and 50–75 mm (2–3 in) deep (whatever dimensions suit your garden fork). Fill this void with loose friable animal manure, or other primary food source, to a depth of approximately 100–125 mm (4–5 in), making the filling substantially higher than the existing bedding as it will diminish in size due to compaction and being eaten. Gently water the gully bait, being careful not to compact it. After this initial watering to settle the contents, resume normal watering.

Once you have the worms trapped and happy in the bait, collecting them is a simple case of using a wheelbarrow and a garden fork. Slide the fork just under the (now compact) baited layer and toss into your wheelbarrow—continuing in a forward direction until the perimeter is completely cleared.

You now have your worms and an empty gully ready to be set up again.

Gully-baiting This bed is full of castings and demonstrates gully-baiting, where only the edges of the bed are constantly fed and watered. The entire bed, however, must be kept moist.

This system works exceptionally well with outside beds exposed to the elements and should be used in conjunction with the shadecloth harvesting technique.

Packaging

The worms must be packed and transported as quickly as possible after harvesting.

Packaging worms in the summer should preferably be done in the early hours of the morning or late in the evening and out of the sunlight. On extremely hot days when worms are frothing badly and are hot to touch (when down to raw form) you may have to immerse them in cool water to stabilise them prior to weighing and mixing them in their packaging material. Simply drop the lot in a bucket of water for approximately 5 minutes. The water must be cooler than the bed they came from. Gently tip the cooled worms into a calico bag for 5 minutes and hang, or sit the bag on a piece of mesh, to allow drainage of all water. Empty the worms onto the table, weigh quickly and mix into the packaging material.

Packaging materials

Your choice of packaging material simply has to be right, with no leeway for error. This is the end result of your months of labour and a mistake here will cost you money. A large concentration of worms in a confined space will survive only where oxygen, moisture and temperature are controlled.

- Worms en masse have a huge oxygen requirement. They can stress, lose body weight and die if conditions are not right.
- Moist conditions are essential, because the worms will give up their moisture/body weight to a drier environment.
- Worms generate a lot of heat when massed together.
- If the packaging material is too dry the worms will ball up to conserve moisture, generate a lot of heat and probably die.
- If the packaging material is too wet or moisture-retentive, and there is no provision for oxygen to enter it, waste gases cannot escape and are dissolved in the wet packaging material. Death can occur due to the combined effects of poisoning, cooking and drowning.

Cocopeat, castings, calico bags and boxes of waxed cardboard are all you need. Cocopeat (a renewable source) is a superb packaging material that drains and retains moisture at a level that is neither to wet nor too dry, and because of its fibrous nature holds a lot of oxygen. Its springy, non-compacting nature with a pH level about 6 makes it an excellent choice, particularly for long travel.

Castings, the worms' natural environment, are a great choice as a packaging material for short travel, but not for long distances. They retain 6–10 times their own weight in water but have a tendency to compact and heat up, thus expelling oxygen and creating a stress situation over distance.

These two products combined (two parts cocopeat to one part castings) make an excellent packaging/transporting medium. The cocopeat retains oxygen and doesn't compact, while the castings retain moisture and provide the worms with the security of their existing environment.

The cocopeat, which comes in compressed block form, should be well broken up and soaked to reconstitute it and to remove excessive salt (some cocopeats can have a salt content of 6–8 per cent which is deadly to worms, so check its country of origin or ask for a salt analysis from your supplier), then placed on a sieve to drain to its own level. If in doubt about the salt content, wash the cocopeat to be on the safe side.

The castings should be moist and forced through a 10–12 mm (0.5 in) sieve so that it is granular in appearance.

Mix the two products thoroughly; when squeezed, a handful should squelch and bounce back once released. No water should drain out of it to sit in the bottom of the packing container.

How to pack the worms

Do not supply food in the package—moisture and oxygen are all that is required. Putrefying and fermenting foods within a confined space, with no ventilation and a limited amount of oxygen, sets the scene for certain disaster. The packaging material is simply a stable product to get the worms from one place to another with the minimum amount of stress. Once the worms are in raw form and weighed, place them on the packaging material and mix gently so that they are all separated and gritty. They can now be placed in their transport containers.

The quantity of packaging is very important:

- 1 kg worms require approximately 3–4 litres of bulk (2 lb / 1 gal)
- 4 kg worms require approximately 10–12 litres of bulk (9 lb / 2–3 gal).

Worms are very adaptable and will 'size themselves' to their environment, that is, reduce themselves in size to suit the oxygen supply, food supply, moisture content or available space. Even when packed as well as possible in the cocopeat and casting mix, they will suffer from crowding (too many worms in a confined space), but should not lose more than 10 per cent of their body weight up to 4 or 5 days after being packed. After 5 to 10 days, and as long as optimum oxygen and moisture are maintained, they may lose 30–40 per cent in total of their original weight, but still be in condition to survive. Once dropped into a bed with space, oxygen, water and food they will regain their condition very quickly, generally in 2–3 days.

Transportation

For long distance travel it is advisable to pack your worms in calico bags placed on slats for good air circulation. The bags can be watered or rained on without fear of the contents drowning as they drain easily. Calico is porous and allows the contents to breathe—the bags should be closely sewn on the seams and sealed with a rubber band to prevent any worms escaping. Wind blowing around a moist calico bag gives you the water-bag or 'Coolgardie fridge' effect, keeping the contents cool even on the hottest of days.

Travel means vibration and compaction which drives out the oxygen from the castings (this is what makes cocopeat, which doesn't compact, such a superb packing material).

For short travel any suitable holding container will do.

For long distances in a truck, the calico bags must be placed in a strong, moisture-resistant, insulating box (polystyrene or waxed cardboard). The lid should be sealed with masking tape and a generous number of holes punched in the lid and around the sides to allow the contents to breathe.

Avoid stacking bags directly one on top of the other, and use your common sense about the number of bags to a box.

Never try to transport large quantities of worms any distance in bulk, that is, dumped out of a bed with a front-end loader into a truck or trailer. Once the vehicle begins to move, the vibration and compaction pushes the oxygen out and seals the worm into his own grave. Going around the corner is okay but any further than that will be a disaster.

CHAPTER 9

Castings

Worm castings are simply worm manure, a readily assimilable smorgasbord of plant nutrients—the by-product of worms consuming food.

The quality of the castings depends entirely upon the quality of the food being consumed—the old saying, 'rubbish in, rubbish out', applies just as much to castings as to anything else.

The value of castings to plants originates in the genetics of millions of years of evolution, resulting in a direct symbiotic relationship between worms and plants. If you have healthy growing plants then you probably have worms in your soil and vice versa.

Castings are a stable product with a carbon/nitrogen ratio of less than 10:1 that does not eat into itself or compost any further.

Good quality castings:

- Have an earthy non-offensive smell
- Readily dissolve in water and can be used as a liquid fertiliser
- Retain many times their own weight in water.

Castings are unique in their shape and design, which allows a very large surface area for the presentation of nutrients. Other then their nutritional value to plants, castings exhibit a micro-organism population which is unique to the worm and essential for contribution to a healthy biomass.

Potential markets

The castings market is an emerging entity with enormous potential in the organic plant food industry. Success in the market place depends upon consistency of the product and documented analysis of trials, application rates and performance on a wide variety of plants from an accredited industry body.

Bed management

Castings, theoretically, are all the product below the working layer. Castings can be divided into good quality (vermicast) and poor quality (vermicompost). Overfeeding and mismanagement can mean that you end up with vermicompost rather than vermicast. Pure vermicast is 100 per cent eaten out and this, of course, is an impractical aim. With a suitable feeding and stocking density management plan, you will end up with good quality castings which for all intents and purposes will be an acceptable stable product.

Castings should be taken from the grower or fattening beds where you are feeding consistent good quality foods. The high stocking density means this food is eaten out very quickly and the unfed areas are also being worked over all the time.

The texture and feel of castings varies considerably depending on the primary food source. Castings produced from the manures of grazing animals which digest a lot of grit (sheep, cattle, horses) is different from castings produced from the manures of animals fed processed food (pigs), which is different again from the castings of worms which are fed fruit and vegetables.

Harvesting castings

Castings can be removed from the top or bottom of the casting layer, depending on the size of the bed. Dig the top over and use the light to force the worms to go down. Then physically scrape or rake the top layer of castings off.

To take castings from the bottom, remove the top 100–150 mm (4–6 in) with the worms and place into another grower bed, then take out all the material remaining. This is the simpler method.

Drying castings

Castings should be drained by whatever method suits your scale of operation. Depending on the quantity, e.g. one tonne or a bucketful, castings can be drained on a clean (no weeds) patch of earth or on shadecloth laid on the soil and left for a few days. In a shed is preferable. The castings may have to be tossed from time to time to assist the drying process. The required moisture content is indicated by a squeeze test—pick up a handful, squeeze tight and then release. The castings should hold shape, but when massaged should crumble apart, neither staying stuck together like plasticine or putty (too wet) nor disintegrating into dust (too dry). Once they reach the required moisture content, they should be covered to retain structure and moisture until packaged.

Packaging and storage

- Castings packed too wet will turn anaerobic and go mouldy

- Castings packed too dry are very difficult to re-wet
- Castings of too fine a structure will turn to a slurry when wet
- A mixture of aggregate size particles that will pass through a 10 mm (0.4 in) sieve is optimal
- Castings packed and sealed in a plastic bag with a couple of pinprick holes, with the right moisture content and structure, will have a long shelf life.

Weed seeds

Quality castings should be free of weed seeds. This is an area that presents a real problem, particularly when you feed manures from grazing animals, or feed fruits and vegetables. Worms will not eat seeds and sooner or later a large percentage of them will germinate. Many seeds, e.g. clover and rye grass, are resistant to the harshest of conditions (stomach acids and partial composting) and will seemingly germinate no matter what you do to the manure. The real problem emerges when moist untreated castings containing a particular weed seed are transported from one part of the country to another where that weed does not exist.

Complete composting of manures beyond the primary thermophilic stage will drastically reduce the germination rate.

Using pig manure or poultry layer manure eliminates the weed seed problem because everything in these creatures' rations is hammer milled before it is subjected to the heating and sterilisation procedures of granulation or pelletisation.

Liquid castings

Where beds are set up with drainage systems for the collection of run-off, it can be used straight away as a liquid organic plant food. This is not necessarily liquid castings.

Depending upon the stage of the bed's development the major part of this leachate may be washed raw manure—which will be strong in odour, rich in colour and is really no more than raw sewage.

This product, if you do not choose to pour it on your garden, would be better employed by recirculating it over the beds without alteration. Otherwise, you can enhance its properties at the collection point by installing an aerobic digester and then recirculate it over the beds.

The easiest and most logical way to get liquid castings is to dissolve pure eaten-out castings in water (using a cement mixer); you then pour this product through a fine sieve to remove the large colloidal material or whatever has not been dissolved. The end product will display soluble and very fine particles in suspension, the amount of particle matter depending on the fineness of the sieve; the depth of its golden brownish colour will depend on the dilution ratio of water to casting. The richness of the castings will also affect the intensity of colour.

Presenting castings to plants in this form allows the rootlets and rootlet hairs to quickly assimilate the readily available nutrients. Pure worm castings dissolve very readily. The strength of the mix for consistency should be calculated on a water to castings dilution ratio of, for example, three litres of water to one litre of castings. Your final choice of dilution ratio is dependent on trial and error to achieve the colour and presentation (amount of suspended material) that you are happy with.

Castings, and therefore liquid castings, contain live micro-organisms that are aerobic at the time they are sealed into a container. After a period of time (depending upon temperature), the oxygen in the solution will be depleted and the environment can turn anaerobic—the product can literally smell like a septic system. If this happens, leave the lid off and let the liquid breathe until the solution is biologically stable, i.e. eaten out.

CHAPTER 10

Performance Indicators (Breeding)

What do you look for as indicators that your operation is performing to maximum breeding potential?

- *Egg numbers*—should be proportional to the number of worms in the bed. The production of eggs is one of the performance indicators which tells you that the worms are breeding. Although you may not actually see the worms knotted, the eggs will tell the story.
- *Egg colour*—the colour of the egg is an indicator of age; the lighter the colour the more recently the egg was laid.
- *Knotted worms* (mating)—you may not always observe this behaviour as a breeding flush is easily missed.
- *Adolescent numbers*—a mixture of youngsters of various ages indicates continual hatchings.
- *Activity level*—this refers to speed of movement and vitality; active worms are healthy worms, which means that bedding structure, moisture content and aeration are generally okay.
- *Health and condition*—the colour, sheen and size are important; large worms with big glossy clitella indicate that breeding should be occurring.

Factors affecting performance

- *Structure*—bedding must be loose and friable with a sealed surface to allow worms to move freely and operate at their own level.
- *Biological life*—a variety of insects and micro-organisms indicates a living environment (the biomass); a bed devoid of insects is regarded as stale and in need of rejuvenation.
- *Temperature*—variation of temperature is important for breeding; when the ambient temperature is right, worms will be triggered to breed.
- *pH*—a neutral pH, which swings in favour of alkalinity, is optimal.
- *Moisture*—not too wet and not too dry; worms are soft, moist, sensitive creatures with a mucous layer that must be maintained—but they are not fish.

- *Aeration*—maintained by regularly tossing the bed to a depth of 100–150 mm (4–6 in) to introduce oxygen which has three functions: the worms breathe it through their skin, it is essential to keep the working layer aerobic for insects and micro-organisms, and it insulates the working layer.
- *Stocking density*—this is a relative term and is subject to your management techniques; understocking is okay, but overstocking will greatly reduce breeding rates.
- *Surface area*—relative to stocking density; worms need room to move and every creature has its space requirements; important for interchanges of gases, evaporation, surface heating, etc.
- *Bed volume*—depth is essential for drainage, moisture retention, temperature variation, micro-organic activity, etc. and is important for maintaining and holding the ambient temperature.
- *Food supply*—worms have two basic instincts, to eat and reproduce, both of which they do very well; a balance must be struck so that both can occur; a rising plane of nutrition is important for maximum fertility and ovulation.

Temperature

The single greatest influence on breeding is variation in temperature; a constant temperature will not give results. Because I rely on the tried and true finger test rather than a thermometer, I do not know what precise temperature range I work with, but it is more than likely 20–25°C (70–80°F). However, temperature is only one of the factors in Nature's Equation; important though it is, you will find worms reproducing in temperatures outside this range because other factors such as pH, aeration, stocking density, food supply, microbes, moisture and volume, will override it.

Spring is, of course, nature's time, when everything will reproduce; the ideal worm breeding conditions are what I refer to as the two-blanket night followed by a warm sunny T-shirt day.

This is the time when the night-time bed temperature drops (for argument's sake) below 20°C (70°F); as morning comes, the temperature rises slowly (this is important) until it reaches what I refer to as the 'ambient temperature' (the ideal temperature for reproduction which must be held for as long as possible). This is the temperature range where all reproduction will take place. Once the bed temperature rises another couple of degrees, breeding will cease for the day.

I have never observed worms breeding under a reversed temperature movement, from hot to cold. They always respond to a temperature moving from cold to hot, like coming out of winter into spring—worms are not intelligent creatures and can be stooged every 24 hours under artificial or controlled conditions into believing it is spring.

In spring all the rules fly out the window—no matter what you do right or what you do wrong, your worms will breed.

Temperature:

- Influences the type of microbes that will flourish
- Affects the pH
- Determines the water-holding capacity of the bedding
- Is the catalyst for chemical changes in the bed
- Has a direct effect upon insect life
- Is the single most important factor synchronising all the variable conditions that constitute Nature's Equation and results in a breeding flush.

Adjusting temperature to encourage breeding

Worm growers basically rely upon the cool of the night to drop the temperature and the warmth of day to raise it. Whatever husbandry practices you employ should be designed to complement nature's thermostat.

- Covers on or covers off to trap the heat in or let the heat out
- Plastic on or plastic off to magnify or absorb heat
- Tossing or aerating to fuel the fire of decomposition or create an insulation layer
- Watering late in the evening to cool the bed (warmer months)
- Not watering in the day to allow extra warmth in the bed (cooler months)
- Decreasing the volume in the bed so the temperature variation will happen quickly (warmer months)
- Increasing the volume of the bed so it maintains and holds the ambient temperature longer (cooler months)
- Increasing the fibre content of the bed to trap more air and begin the composting process or to insulate the top against heat or cold
- Using fermenting foods within the bed to generate heat via microbes and composting
- Using freeze-dried microbes, fibre and molasses to generate heat
- Elevating beds off the ground to allow air to flow over and under the beds for cooling
- Heating cables in the beds in the cooler months
- Protection against the wind to reduce the wind chill factor in the cooler months
- Shadecloth or covering to protect against harsh direct sunlight

- Positioning beds to receive the maximum advantage of heating and cooling in all seasons.

The many things you can do to influence temperature and affect breeding should be done regularly, not on a stop-start basis. It takes time to encourage and maintain the conditions in the bed that reflect spring and encourage breeding.

Clitellum

The presence of the clitellum, the band or protrusion ('saddle') behind the head of the worm, is an indication that the worm is sexually mature. Essential for the mating act, on a sexually mature worm the clitellum is always present but is not necessarily pronounced, coming and going with the condition of the worm. The enlarged clitellum is a sign of a worm's 'coming into heat', primed hormonally for the act of reproduction.

A bed of primed worms in good condition with glossy skins and pronounced clitella should be breeding. If they are not, then usually either pH or temperature needs attention.

Temperature and incubation

The worm will instinctively lay its egg where it has the best chance of survival at that particular time. In a covered bed the obvious place is just under the cover where the egg will be able to absorb the heat from the sun and yet remain moist.

Incubation requires particular temperature and moisture conditions. A higher constant temperature (lukewarm to touch) is required for incubation than for breeding. It is important to have eggs hatch as quickly as possible after they are laid. The egg capsule is a semi-permeable membrane with from 1–20 embryos encased within its walls. There is only a limited amount of sustenance within this capsule and once it is depleted, some embryos die so that others can live—survival of the fittest. Capsules laid in autumn without a sufficient incubation temperature will survive until spring but you will only get one large worm emerging for each egg.

An incredible number of eggs can be achieved from a 21-day cycle (see page 34), but due to poor incubation conditions (mainly excessive heat), the potential hatchlings can have a devastating mortality rate leading to very poor results.

Survival management in temperature extremes

Heat

In extremely harsh summer conditions where bed temperatures far exceed any possibility of breeding, it is advisable to simply lock up the beds. Enter summer with good depth of casting in the beds below the working layer, and simply mulch the top to a depth of, say, 25–50 mm (1–2 in). Mulch provides the best protection against direct heat from the sun or radiant heat from a shed roof, and can be watered through to keep the bedding and the worms moist. They will not breed and they do not need to be fed. This is survival management. The worms will size themselves down to suit their environment but they will survive, escaping the heat by burying themselves in their own cooler castings.

Replace the mulch layer only as it breaks down through natural decomposition. When the harsh weeks or months are over and the temperature is more amiable you can begin to restore the bed by working on the top 150 mm (6 in) and implementing normal worm husbandry practices. The worms will still be there. They will bounce back to health, condition and size very quickly.

Cold

In extremely cold conditions you have one of two choices: generate heat through the composting of manures, microbes and fibre in the beds; or lock the beds up for winter—it is too cold to breed anyhow, so why feed them.

Maintenance is much the same as for extreme heat conditions. Enter winter with a degree of depth in the beds but with a generous amount of fibre that will decay and provide some sort of food. Apply a generous amount of mulch over the surface to keep the frosts and heavy dews off the bed surface—basically a blanket against extreme cold. Watering is kept to a bare minimum and the mulch layer replaced as it breaks down through natural decomposition. The worms will survive by sizing themselves to their environment and will quickly bounce back to health and condition once normal worm husbandry practices are restored.

pH levels

The term pH refers to the level of alkalinity or acidity present as measured on a scale from 1–14, where 7 is neutral.

Controlling pH is one of the most important aspects of worm breeding.

The pH level in spring is not particularly important because all the other components that affect breeding exist in the right proportions and can override unfavourable pH levels. No matter what you do right or what you do wrong in spring, your worms will breed. However, once the spring flush is over it is of paramount importance that you correct the pH because the other components of Nature's Equation begin to run out of balance and you have to jiggle it back within specific boundaries.

Adjusting pH levels

pH is easy to adjust. Your scale of economy, the economics of the solution and its manageability will determine how you go about it.

Peatmoss (around pH4) Peatmoss is a non-renewable source and a valuable commodity in the agricultural and horticultural industries. It is a food source for the worms of non-protein origin and retains moisture very well. I find it absolutely excellent to use. It comes in a large compressed sealed bag, so storage is easy. Its structure is very fine and it blends very easily with whichever manures you like to use. Its advantages far outweigh its cost. Peatmoss can be blended with manure prior to application or added on top of the bed and forked in afterwards. The breeding beds require only small volumes of material so adjusting the pH with peatmoss is not a costly exercise to achieve the end result.

As a rough guide, one part peat moss to two parts manure by volume is a fine starting point, subject to the source and pH level of the manure. The result can only be tested a few days down the track, when the processes of chemical interchange, worm digestion and aeration, watering and so on have settled.

Acid For the larger operation with greater expanses of surface area, it is more practical to use something like hydrochloric acid—the same that is used in swimming pools. As an example, to a 1000-litre *plastic* tank of water you add maybe two litres of hydrochloric acid. Test the pH with the appropriate water pH testing kit until you achieve a reading of about 5.5–6.0. Every time the beds are watered with this solution you will be subtly changing the pH. See aerobic digestor, page 20.

Supplementary foods Implementing a programme of dollop-feeding fermenting foods such as pollard, or putrefying foods (vegetables and fruit), will create localised acidic areas. Regularly tossing the bed ensures even distribution of these acidic patches with alkaline manure patches and a balanced neutral bed can be the result. Although worms are popularly supposed not to care for citrus, they can eat anything. I can assure you that there is nothing wrong with using acidic foods, such as oranges and their by-products, when your aim is to adjust pH. (However, large quantities of acidic foods should not be fed constantly.)

pH testing

To do a pH test you must take a representative sample of content from the working layer. Take 10 or 20 handfuls from all over the bed, place in a wheelbarrow, bucket or cement mixer and toss for even distribution before taking a sample or two for testing.

The best tester to use is one designed as a soil-testing unit and endorsed by an accredited agricultural industry body, e.g. CSIRO or departments of agriculture. A colour metric tester which uses a colour card for easy readability is quick, economical and easy to use.

The colour will show you the status of your bed and it is then up to you to adjust the situation as you see fit.

To lime or not to lime

Lime is calcium carbonate with a pH reading of 9.5, so it is very alkaline. It does not make any sense to use lime if your beds are already alkaline. Worms are invertebrates and do not have teeth and bones; their requirements for calcium are met in their regular food sources.

When would you lime? If your food source is acidic or turning acidic via fermentation, e.g. fruits and vegetables and flour products, then there would be a reason to use some product to neutralise the bed for breeding purposes. Lime (preferably in the form of dolomite, but agricultural lime is okay), or the use of animal manures, would rectify the situation.

pH in growing beds

Until the castings produced under your system become cost effective, there is no need to go to the expense of adjusting the pH in growing beds, as the worms don't breed in their conditions of high stocking density. When castings become worth selling, then it is necessary to standardise the pH of the beds at around 7 to satisfy plant nutrient requirements.

pH and nutrients

As you change the pH from alkaline to neutral the working structure of your bed changes. The bed becomes loose, friable and workable as it frees up nutrients previously locked up, thus giving the worms food that was previously unobtainable.

Aeration (tossing)

This is an essential part of normal worm husbandry practice because:

- Worms are aerobic creatures
- Microbes in the working layer must also be aerobic
- Air provides oxygen
- Air is your insulation for temperature control
- Tossing gives bedding a workable structure.

When do you toss your beds?

- On a regular basis, every few days
- When compaction occurs
- When they become waterlogged. Poor drainage will necessitate constant tossing to combat anaerobic conditions

- When they are too hot or too cold
- When uneaten food must be brought to the surface—this probably means you are overfeeding
- When excessive fibre in the bed requires constant tossing to allow the injection of oxygen to fuel the fire of decomposition to generate heat and break the fibre down
- When heavy downpours of rain force oxygen out of the bedding and the worms must come to the surface to be able to breathe (this will precipitate a walkout).

How to toss

A fine pronged fork such as a pitchfork is essential to minimise damage to worms while aerating. Hold and insert the fork at a 45° angle and use a lift, twist and roll action.

Under normal husbandry conditions, only toss to a depth of 100–150 mm (4–6 in) (working layer). This brings old castings up into the new food, which is good as it blends and mixes essential microbes, chemicals and enzymes to assist breakdown and create a more desirable working environment for the worms.

What happens to the worms when you toss?

When the top 100–150 mm (4–6 in) has been tossed and the worms have been disturbed, you will find that the majority, for a short period of time, will go to the bottom of the layer. They then begin to work and eat their way up again, progressing as fast as the oxygen supply runs out.

Worms in high stocking density numbers are efficient vibrators and compact everything under them. The time frame required before aerating again is totally dependent upon how quickly this progressive compaction, or settling action, eliminates oxygen from this layer. This is why it is important that your working layer has the correct non-compacting structural components (inert fibre) to capitalise on friability, workability and oxygen retention.

Always remember that worms size themselves to their environment (like goldfish in a bowl)—this means that if there is little oxygen in the bed they very quickly reduce in size to suit the availability of oxygen—the same with food and water.

Healthy, well-conditioned worms are essential for reproduction and growth—to starve your stock of oxygen is to starve yourself of a return.

Oxygen and microbes

Worms live off the soft decaying tissues and by-products of microbial decomposition, and the microbes themselves. Worms do not have teeth.

They have very little mouths and project their intestines out of their mouths to secure food.

Oxygen is essential for aerobic microbes, which you must encourage. Microbes divide by mitosis every two hours—so an enormous population of microbes can be encouraged in a very short period of time.

As long as the primary food source and moisture and temperature are present, oxygen is the catalyst for a food source practically immediately available to the worm. Fresh and semi-fresh manures are packed with millions of microbes. Once the microbes are placed in the bed it is not hard to encourage their proliferation.

Aeration and insulation

The insulating effect of fibre bats in ceilings and walls comes not from the fibres but from air trapped within the fibres. The same is true of a doona.

If the worm beds are hotter than the outside temperature they must be tossed to let the heat out; keep tossing and watering until they cool off.

In winter, if the beds are cold, try tossing them to trap air within the bedding and then cover with black plastic or bubble wrap to absorb and magnify the heat generated by the sun. As the day cools off replace or cover the plastic (depending on the severity of the temperature drop) with an insulating layer of bags, carpet or underlay to trap the heat and prevent it from escaping.

Aeration and supplementary feeding

It is absolutely essential to aerate after having used fermenting or putrefying foods, particularly those sourced with a lot of liquid which drains into the working layer of the beds. These foods should be restricted to narrow sections of the bed by strip feeding. Once the strips of food are practically eaten out the areas beneath them should be tossed and watered, and checked the next day for heat generation and smell. When any smell and/or heat has dissipated it is safe to feed in this area again.

A lot of supplementary food gets carried or vibrated below the surface. An accumulation of uneaten food can poison sections of the beds—so after a flush of protein food, make it a husbandry practice to toss your beds.

Bedding volume

The relationship between bedding volume, surface area, drainage, and evaporation is important for the retention of temperature and moisture.

In breeding beds, the depth or volume should suit the season, being kept shallow in summer to allow quicker cooling and to prevent composting taking place, deeper in the winter to allow for composting and the retention of heat.

Breeding beds should be coned or domed, say 150–225 mm (6–9 in) in the middle and tapered down at the sides to 50 mm (2 in). This shape allows for variations in temperature and drainage within the bed, being warmer in the middle and cooler on the outside, wetter in the middle and drier on the outside.

The volume must be sufficient to hold the ambient temperature for long enough to encourage breeding, with the capacity to raise the temperature above the ambient temperature and likewise drop below it, all in a 24-hour cycle. If the bed is too deep this all-important temperature fluctuation will not occur.

In growing beds the volume does not matter and the only factor to be concerned about is drainage. As long as your feeding practices do not result in a huge composting bed, your only restriction will be the management of a huge volume at clean-out time. The deeper the bed the more compacted are the castings below the working layer, which may cause drainage problems. You may find that you have to water more frequently, in smaller amounts, to keep the system operating. The water will take longer to filter down and you may think you are flooding the beds. You are not—you merely have a slowdown in filtering because of the volume.

Bedding structure

Bedding structure must be loose and friable from both the worm husbandry and management points of view. An easy worked bed is an easy managed bed.

Structure will depend on correct pH, the quantity of cellulose and stocking density.

A loose and friable bedding structure retains moisture, retains oxygen, encourages the proliferation of microbes and insect life and is a healthy working environment for the worms.

In growing beds, the massed numbers of worms should keep the surface 50 mm (2 in) bubbling as long as you jiggle the cellulose and protein requirements just right. Everything below this working layer should be semi-compacted because of the vibratory nature of the worm yet should still facilitate good drainage. This castings layer should be practically devoid of cellulose/fibre.

In breeding beds, with a lesser stocking density, the worms should be found all the way through the bedding which shouldn't be deeper than 150–225 mm (6–9 in) in the summer months.

CHAPTER 11

Worm Behaviour

The smell of stress

Worms, like any other creature, have their own smell. You can quickly establish their state of distress by the intensity of their odour. It is normal husbandry practice to be constantly sniffing the beds. It is important to observe your worms' behaviour and make an assessment of their environment in response to any smell of stress, which is a symptom of a condition.

There are four levels of odour:

1 *The healthy bouquet*

The normal smell of worms in a healthy environment is earthy with a hint of worm.

2 *The distinctly wormy smell*

This smell is indicative of worms in need of moisture. Their scent overpowers the medium that they are in—the worms are usually quite active but may exhibit signs of the 'survival mode' (balling up or looking for an escape route). They are losing condition and body weight but can be revived at this level. At this level of distress, they still maintain their outer moist layer.

3 *The dying and the damned*

At this level of distress, the worms have a pungent, acrid, clinging, absorbing smell that screams at you—the worms have lost a lot of condition and are exuding a yellow-green oil. Their movements mirror their pain; they are disoriented and obviously very distressed. This smell is common while harvesting in unsuitable heat. At this level, some will live and some will die.

Immersion in cool water or saturated cool bedding is their only chance of survival. This is the level where the worm cannot maintain its outer moist layer.

4 *The smell of death*

Totally gagging and overpowering. Once smelt, never forgotten.

'Gut-shot' syndrome

To be 'gut-shot', in my terminology, is to be shot through the abdominal region—this spreads foreign material and micro-organisms into the abdominal cavity, which is the precursor to peritonitis, a slow and horribly painful death unless treated very quickly with massive doses of antibiotics and surgery.

This scenario in your worm beds is very painful to see. It is usually too late once you realise what has happened and you more than likely lose every worm in your bed.

Symptoms The syndrome usually starts with a considerable number of worms floundering around disorientated on the surface of a bed. Normal reaction is probably to toss and water the bed and adopt the 'wait and see' principle. The next day there will literally be tens of thousands of worms dead and dying on and in the surface layer. At this point it is practically too late, for if 'peritonitis' has set in, by the next day they will all be dead.

Causes One of the causes is over-feeding with fermenting protein foods, particularly in the heat. With constant overfeeding or poor supplementary feeding techniques (e.g. feeding dry, loose feed and constantly watering it in), you end up with a working layer full of fermenting and putrefying sour foods that the worms cannot keep up with. The result is an accumulation of poisonous gases that also dissolve and are leached down through the bed. Putrefying worms and the gases they produce further aggravate the situation; with the coincidence of one hot day you have reached 'critical mass' and sustained the gut-shot effect.

Treatment Once you realise you have a poisoned bed, act quickly by laying down strips of moist neutral material (no active food value), such as cocopeat, peatmoss, castings, to provide an escape route and safety zone for the ones that will survive. Do not blanket the surface but allow enough space between the strips to let the poisonous gases out. The situation in the bed once it reaches 'critical mass' will not improve and must 'eat itself' or 'burn itself' out. In the very short time that you have, collect as many worms as you can in these survival baits and place in another bed. Once you have removed all the living worms, you can assist the 'burn out' of the poisoned bed via composting (watering and tossing) or simply remove its contents and set up afresh.

Conclusion The gut-shot effect can also occur in overstocked beds, where through overheating the worms begin to die and the poisonous gases of decomposition very quickly intoxicate and engulf the bed. The gut-shot effect can be easily avoided through sensible management practices, particularly in relation to supplementary feeding.

Walkouts

'Migration', or 'walkout' as it is often referred to, will only occur on cool nights under conditions of rain or 100 per cent humidity in the air.

Walkouts can be very common in the cooler months and sometimes may seem not to occur at all—then you will get that special night that will trigger a mass exodus. On these nights it does not matter if all the conditions in your bed are right—the worms will just up and go.

Moisture is the only method that nature has allowed to facilitate these creatures' relocation to another food supply. There has to be a very high percentage of humidity in the air so the worms' systems can remain stable and not dry out, and the surrounds have to be wet so they can move over the surface.

Worms on a walkout do not burrow into the soil for survival but seek out something to bed under, e.g. rocks, bricks, timber, foliage, clumps of grass. They will wander with no apparent direction all night until light of day and then it's simply a case of musical chairs—the music stops (the sun's up) and where they are, they stay. If you have the Aussie blue, you can find them the next day massed 20–30 metres away from the bed. In harsher climates these creatures ball together to conserve moisture to survive and will stay like this until the next night where, if conditions are right for migration, they will set out again. If conditions are not right, they simply desiccate and die. The solitary worm on a barren landscape does not last very long at all—the sun and the wind and the ants quickly snuff out its life.

In greener pastured areas like the coastal fringe with all the kikuyu and paspalum around, the wandering worm will survive because of the ground cover until the harshness of summer arrives. Once the worm puts out a distress signal it is easy prey for the ants which will attack unmercifully—yet in its natural environment the ant and worm co-exist.

So the compost worm, once out of its rich organic environment, is living on borrowed time, unless by a fluke of nature its aimless wandering leads it to another sizeable food source to survive in.

Walkouts from indoor beds

When it is evident that the night will be suitable for walkouts it is important that the lights stay on, that the beds are tossed and moist and have feed. Do the best you can do to compensate for this primitive instinct. See Plate 18.

What is a prime night?

A prime night can be the crisp, cool, still foggy night of winter just after the gentle rain has stopped. Or it can be the balmy still night of spring and autumn with gentle rain.

Walkouts from outside beds (including windrows)

The beds should be designed so that some sort of external covering can be easily applied to prevent rain saturating the beds and driving out the oxygen. When this happens your stock has no choice but to come to the surface and once on the damp surface they are stimulated to begin migratory behaviour.

It may not be practical to have covers over windrows. In this case your only safeguard other than lights is to minimise stocking densities so that only a minimal amount leave. Design your windrows and manage them with walkouts in mind.

External beds that are not covered should be deeply tossed but not watered prior to the onset of a couple of days of rain; every day it rains they should be tossed to inject oxygen into the bed and relocate the worms which have congregated around the perimeter.

External beds designed according to the principles outlined on page 16 should prevent losses of worms that are on the move. Even without lights, good bed design, good husbandry practices incorporating covers, aeration, control of bedding structure, good drainage and safe stocking densities will safeguard against the loss of worms.

Overstocking is the main factor in the walkout equation

In overstocked beds breeding will stop. To compensate for this, the worms' primal instincts compel them to vacate the bed through migration, thus reducing the stocking density, re-establishing breeding and perpetuation of the species.

We artificially induce overstocking for management reasons and compensate the situation with good quality food and husbandry practices to facilitate their health and condition.

Australian native worms are seen after heavy rain all over lawns, concrete paths and bitumen roadways because their burrows are full of water and they have been forced out in search of air. They wander aimless and disoriented, at the mercy of birds and other predators.

Trapping the walkees

Manure traps can be placed at regular intervals around beds where walkouts have occurred to attract the escape artists. The traps are kept moist and sealed with a cover and removed a couple of days later. Prevention is far easier than cure, however—I would tend to regard the odd walkout as experience and an acceptable loss.

Breeder burn-out

This phenomenon will happen with any producing livestock. It happens with laying poultry in cages—when in full production, they may be up to

98 per cent lay. After 12 months they could be down to 50–60 per cent lay, where it is uneconomical to keep producing, so the stock is replaced with point-of-lay pullets which very quickly reach full production.

Milking cows are 'dried off' once they drop to a certain level of production and then go through a resting period until they calve and the cycle begins again.

With worms, nature's built-in resting periods are the seasons of summer and winter. They breed like rabbits in spring, rest in summer, breed in autumn, and then rest again in winter. By providing the right conditions, we should get from the worms approximately three months lay at any period of time before they rest, allowing for a period for acclimatisation, a period of maximum lay and a tapering down to not laying—this is their innate instinct.

The burn-out factor is very real but can be managed simply by rotating and resting your breeding stock or pacing their breeding by altering conditions to provide a rising plane of nutrition. Burn-out also has a lot to do with food supply. To manage a breeding bed efficiently you have to limit the amount of food to complete a cycle of events.

A breeding bed which has become burnt out or stale must be rejuvenated or replaced. When the worms are not burnt out, but the bed is, make a clear distinction.

CHAPTER 12

Problems and Pests

Tree roots

Tree roots can be absolutely devastating in a worm bed. They create a fibrous mat, drain all the moisture, destroy bedding structure and make tossing and cleaning out the bed a management nightmare.

Particularly bad are the evergreens such as gum trees and pine trees, which have a large fibrous surface rooting structure. It only requires one rootlet to gain entry into a bed for the nightmare to begin. See Plate 19.

Deciduous trees can be as devastating in spring and summer. However, they have a soft rooting system that can be easily broken up every time you toss. In autumn and winter they are not a problem because they are practically in hibernation and are not in need of sustenance.

The importance of summer shade for the beds should not be ignored, however. It is not that hard to plastic-line the beds and incorporate an agricultural pipe to facilitate good drainage or to built an off-ground bed to overcome the tree root problem. Trees should be viewed as only an 'initial set-up' problem. The advantages they provide of shade and shelter far outweigh the additional set-up cost.

A rooting problem can delude you into thinking that the beds are wet all the way through when only the top is saturated; in reality the base of the bed will more than likely be bone dry because of the voracious thirst of the matted rootlets.

Ants

The occasional large ant in the bed is of no consequence. The problem ants are the little black ones that swarm by the thousands—the ones that create their own tracks across the landscape for literally 50-odd metres to source out food—but they are still only of nuisance value. They will colonise under or beside the beds to be close to this wonderful new source of food and moisture. They swarm across the beds scavenging for supplementary foods such as pollard which they take back to their nest—they are not lured there because of the animal manures.

It is unlikely that ants will attack a worm or carry off worm eggs, unless the worm, because it is damaged, displays a distress signal, or the eggs are non-viable following exposure to sun or wind. The eggs carried by ants are more than likely their own.

Ants become a problem when they are so numerous that you cannot sit on the edge of the bed or dig the surface without hundreds of them crawling all over you—besides constantly getting bitten, this is a horrible sensation.

Once you realise a problem is emerging, stop supplementary feeding (to remove the attraction) before the situation becomes critical. You can end up with ant nests and colonies everywhere around the beds, including the spaces between the plastic liner and the timber sides. Once a nest is established you will have to poison bait the ants to wipe out the colony.

Colonising ants can live in harmony with worms as long as the worms' environment is safe. An ant colony under the plastic of a windrow is usually okay. However, if the worms are stressed by exposure to sun and wind (by lifting the plastic), they will be attacked.

From the point of view of aggravation, ants should be controlled. Carefully consider bed location with regard to nests of ants, particularly in the drier regions where box and gum trees favour their existence.

Insect plagues

Usually in spring (when all of nature's beasts proliferate), you can end up with plagues of all sorts of insects, including red mite and springtails. This is not necessarily a bad thing, but it can be aggravating. For there to be a plague of anything, the conditions have to be right and primary to this equation is the food supply.

Controlling an insect infestation is usually as simple as changing the worms' food supply or altering its presentation. Plagues will burn themselves out as other factors (temperature or pH) change due to seasonal conditions. Such plagues do not affect the performance of your livestock and can be overlooked for the short term.

Mice

Mice are really destructive, dirty creatures, doing more damage in your storage shed than to your worm beds. If you cannot seal your shed against mice then everything should be stacked in an exposed structure. Bags of food such as pollard are an easy target, as are hessian and paper products.

Mouse and rat baits work very effectively and should be laid in sensible, safe positions that are attractive to the rodents and out of reach of other animals (and children). Sometimes all you need is a barn cat.

Mice are active foragers and very migratory, so it is not unusual, when there are a few around, to see them scampering over the beds at night looking for food (grain-originated products).

The only time they pose a problem in the worm beds is when they burrow under the covers to set up their nests. They are very ingenious and destructive, using portions of the covering to line their new homes. Mice work very quickly—one day they are not there, and yet the next day it seems there are tunnels everywhere. Solving the problem is as simple as removing the covers and tossing the beds to destroy their nests. If the surface area involved is too large to repeat the tossing every couple of days, then the covers will have to stay off until they stay away.

If you do have to leave the covers off for any time, replace them with mulch or blanket feeding. If the mouse problem is coupled with a bird problem, cover the beds directly with, for example, 70 per cent shadecloth.

If you are blessed with barn owls or tawny frogmouths in your district, you will find them superlative night-hunters of the mouse. The butcher-bird and kookaburra are equally deadly in the early hours and late evening.

An excellent predator of mice at nesting time is the snake. A carpet snake or diamond python living in your shed is a great asset, but unfortunately not everyone is happy about this, especially when you can't be sure whether the tracks you find belong to a friendly type or a poisonous brown, black or tiger snake. This is paranoia you can probably do without.

Birds

These beautiful creatures can appear to be your enemy if you do not initially set your beds up with the bird population in mind. Birds are not the most intelligent of creatures and those attacking the worms are simply following their gastronomic instinct rather than displaying malign intent.

Not all birds are meat-eaters; a lot are insect-eaters. This is not to say that if a worm was on the surface an insect-eater would not peck at it, but by the same token it is not always their instinct to root through the soil looking for tasty morsels. Birds like parrots eat seed and grain and will not go near your beds at all, other than to perch.

Birds like the soldierbird, the peewee and willy wagtail are insect-eaters and their presence is a response to the presence of flies, small spiders and insects on or around the bed. The butcherbird hunts large spiders, large insects, moths and mice.

Birds like the black jay are very destructive and pick and root through anything, making a hell of a mess looking for larvae, worms, grasshoppers, etc. Magpies and kookaburras look for a movement in the soil and will take worms.

The beds can be set up initially with multi-functional external covers of shadecloth or tarpaulin to combat heavy downpours of rain, block direct sunlight and keep out birds. Simple internal covers of hessian bags or underfelt will also keep the birds off. Surface covers of 70 per cent shadecloth will also keep the birds off.

Leaving the beds unprotected and relying on scare tactics does not work. Birds can be a major aggravation and you can shoot them, poison them, catch them, run over them, or bludgeon them to a pulp (all illegal, by the way), and they will still come back. Using covers to fix the situation is not hard and will give you peace of mind.

The meat-eaters will be deprived but the insect-eaters will thrive and you will be amazed at the rapid disappearance of flies, moths, and spiders (there are some nasty ones around) in and around your shed and beds once you encourage these species (simply by not discouraging them).

Maggots

Maggots are the larvae of the flies attracted to putrefying or decaying foods, particularly protein foods. The presence of maggots in any large number in your beds usually coincides with extra hot days and the presence of supplementary foods such as pollard. Maggots look disgusting and conjure up horrible thoughts, but can quite easily be eradicated by removing the source and tossing that section of the bed. For this reason, meat and fat products should be avoided. The presence of an occasional maggot (and other insect life) in the bed should be of no concern, as it simply reflects the presence of food.

A Few Commandments

1. The breeding of worms is about surface area.
2. The breeding of worms is about stocking density.
3. The breeding of worms is about variation of temperature.
4. Do not fall in love with your worms.
5. Come to grips with the notion of acceptable loss.
6. Worms are one of nature's beasts and are subject to the rules of Nature's Equation.
7. Worms do not have an intelligent brain—they react to external stimuli and survive by instinct.
8. Worms are livestock and therefore require good husbandry management skills.
9. Worms are slivers of protein and require good protein supplementation.
10. The only thing that that will kill worms is bad management.
11. Worms will vacate your beds.
12. Match the food supply with the ability of the worm to eat it.

Glossary

Acceptable losses The stock written off through unavoidable inefficiencies in the operation

Adolescents Junior worms prior to sexual maturity

Aeration The injection of oxygen into a system, usually into the working layer, by the process of tossing

Aerobic digestor A vessel to collect leachate from a bed, in which a submersible pump injects oxygen into the liquid and its composition is changed through the action of aerobic microbes

Ambient temperature The specific temperature range required to trigger and maintain a breeding flush

Baiting The process of encouraging worms to congregate at one location by the placement of food

Balling up A survival behaviour in which worms congregate together in a ball to conserve moisture and protect themselves against a hostile environment

Blanket feeding The application of a food source as total coverage of a bed's surface

Breeding flush A period of intense breeding triggered by optimum conditions

Castings (cast, vermicast) Worm manure, the end product of the worm's digestive process

Cocopeat A fibrous product derived from the breakdown of coconut husks, used as a worm packaging material

Dispersal and disorientation A distress and survival behaviour in which the worms wander aimlessly away from or out of a hostile environment, even in broad daylight

Dune-shaped The cross-sectional shape of a worm bed resembling a sand dune in appearance (*also* dome-shaped)

Egg/capsule The structure containing the developing worm embryos

External covering Covers used as protection above, but not on, the surface of the bed

Friable A loose, moist and crumbly structure that is easy to work

Hatching flush A period of intense hatching triggered by the right conditions

Husbandry management The practices, principles and procedures of maintaining stock in good order and condition

Hydrochloric acid A liquid product used in water to reduce the pH of an alkaline environment—toxic in concentrated form, handle with extreme care

Internal covering Cover used as protection on the surface of a bed

Leachate The fluid that drains through the worm bed as a result of watering, which can be collected as run-off if required

Light-extraction The process of harvesting worms by hand, using the principle that worms are photophobic and will move away from a light source

Management system The operational method employed to complete a cycle of events, from breeding through to harvesting

Microbes/micro-organisms Biological life at the microscopic level, essential for a healthy worm environment

Molasses A thick black sugar syrup, a by-product of sugar refining

Nature's Equation The synchronisation of numerous variable conditions at a particular time to result in a breeding flush

Off-ground beds Beds elevated sufficiently to be off the soil surface, to avoid tree roots or facilitate the collection of leachate

On-ground beds Beds positioned on and in contact with the soil surface or very close to it, e.g. on a concrete base

pH A measure of acidity or alkalinity

Pit-run worms The community of worms from a typical multi-function bed, comprising newly hatched, adolescent and sexually mature stock

Pollard A fine white powder, a by-product of milling flour from wheat

Primary food source The main food supply, usually animal manure, that is readily available and economical to use

Raw worms A mass of naked and exposed worms out of their bedding

Secondary food source The structural component of a food supply that supplies additional fibre and also has nutritional value

Stocking density The amount of worms that suit, or can be placed into, a particular unit of space to achieve the management objective; measured in kilograms per square metre or numbers per square metre, not cubic metre

Supplementary food Concentrated food of a protein source that encourages rapid microbial activity

Surface area The area of a bed from a bird's eye view, measured in square metres

Temperature The degree of heat or cold within the matrix of the worm bed

Tossing The process of turning the contents of the working layer with a pitchfork

Vermicompost The intermediate stage in the production of castings, where food material is still being worked over by the worms

Working layer The top 100–150 mm (4–6 in) of the bed that is tossed, aerated and fed, and where 99 per cent of the worms should be located

Index